57:1 · April 2019

ENGLISH LANGUAGE NOTES

Introduction

LAURA WINKIEL

The oceanic turn has been under way for some time now. Whether we date this turn as occurring with the publication of Fernand Braudel's *The Mediterranean and the Mediterranean World in the Age of Phillip II* (1949), with Paul Gilroy's *The Black Atlantic: Modernity and Double Consciousness* (1993), or with oral and written work by Indigenous and other intellectuals from the Caribbean, Oceania, Africa, and elsewhere, this strand of historical and imaginative thought has, in the past two decades, coalesced into a far more complex formation.[1] One major aspect of work that has been woven into humanist and Indigenous accounts of the oceanic is the scientific exploration of the seas, a process that began with the earliest European imperial voyages.[2] In the past few decades oceanography, geophysics, and evolutionary and marine biology have drawn our attention down to the smallest units of the microbial seas, undetected through ordinary human perception, and upward to a scale of millions of years of evolution. The result of this scientific calibration along inhuman scales has been a reconceptualization of what life is and where it occurs. For instance, recent work in cellular biology and the DNA sequencing of marine microbes, as Stefan Helmreich reports, have posited life forces for which we have no conceptual categories: "Aliens [marine microbial realms inaccessible to prosthetic-free human experience] are life forms whose place in our forms of life is yet to be determined."[3] Though often these genetic and microbial discoveries are commodified on behalf of corporations, in the hands of science studies scholars and environmental activists the net effect has been to posit the sea as a material actant in a posthumanist world.[4] Stacy Alaimo has called this relation "transcorporeality" to suggest that we coevolve with the seas and are part of, rather than separate from, the unknown future of the world's becoming.[5] For instance, humans have polluted the seas with heavy metals, nuclear waste, and plastics, with the result that human bodies, along with marine and terrestrial animals, carry profound, often life-threatening, toxic burdens that are assumed unevenly across the globe.[6] In addition, the concept of transcorporeality questions the guise of detached objectivity that science has long claimed and demands self-awareness of how we impact—and mediate—the very environments we study.

The second strand of sea thinking to join with the sciences and humanities is environmentalism, an interdisciplinary mode of inquiry that draws not only on

57:1, April 2019 DOI 10.1215/00138282-7309633

these disciplines but also on the social sciences and activist art and politics. From the incontrovertible evidence that we are in the midst of the sixth extinction to ocean acidification and warming temperatures, rising sea levels, melting ice, overfishing, and oil and nuclear contamination, we are confronted with an ongoing, human-induced catastrophe. The systemic crises of the oceans demand a wide range of responses, from studying oceanic geophysics, to constructing posthuman understandings of the sea and its organisms, to interrogating human/nonhuman marine relations in order to rethink the ethics of human actions toward the nonhuman world so that they might be nondominating, even erotic and pleasurable. Given the sheer multiplicity and stunning possibilities that work on the oceanic brings to paradigms of thought, criticism, visual arts, reading practices, and ethical and political action, "Hydro-criticism" is only one of many recent interventions into maritime critical inquiry. Before I turn to the groundbreaking articles assembled in this special issue, I will briefly highlight important recent interventions in oceanic studies made by scholars working in cultural studies, gender/queer studies, literary studies, and global South/postcolonial studies. The aim of this review is to set the stage for understanding exactly how the contents of this special issue furthers the critical and creative interventions made possible by the maritime turn.

In 2010 a highly visible *PMLA* forum, "Theories and Methodologies," brought attention to the sea not as an "empty" space crosshatched by longitude and latitude markers over which ships traverse but as an ecological environs, a "hydro-commons," and the vanishing point of "nature." Rather than nature, the sea, in the lead columnist Patricia Yaeger's words, has become a "techno-ocean" that "subtracts sea creatures and adds trash" on a sublimely horrendous and ecocidal scale.[7] She memorably includes a photo of a dead baby seabird whose autopsy revealed 12.2 ounces of plastic debris in its stomach that its parents had mistaken for food, resulting in its death of malnutrition and dehydration. In the same collection, Margaret Cohen, a contributor to this special issue, called out literary critics' "hydrophasia," a forgetting of the sea whereby *Robinson Crusoe* becomes a tale of economics rather than shipwreck; *Moby-Dick* concerns factory labor; and Conrad's "The Secret Sharer" plumbs the ship commander's narcissistic personality.[8] Unlike Braudel's or Gilroy's earlier interventions, Cohen's critical emphasis comes squarely on board the ship itself, rather than on how regional geographies (the Mediterranean and the Atlantic) can transform critical paradigms. She stresses the work of navigation and sail and how narratives of maritime mobility and practical reason helped invent the novel. Her maritime criticism, to some extent, parallels the groundbreaking work of the historians Peter Linebaugh and Marcus Rediker, whose attention to impressed sailors, slaves, pirates, and other outlaws of early Atlantic modernity demonstrates the long history of working-class resistance and cross-racial alliance that had been largely erased from view.

Another contributor to this *ELN* special issue, Hester Blum, in the same *PMLA* cluster, notes, "If methodologies of the nation and postnation have been landlocked, . . . then an oceanic turn might allow us to derive new forms of relatedness from the necessarily unbounded examples provided in the maritime world."[9] She charges that concepts such as citizenship, sovereignty, and labor, when viewed from the masthead, take on guises unfamiliar to nation-based assumptions. These counterhistories challenge the homogenizing assumptions of nationalism and anticipate a future—and recover a past—that is multiracial, multilingual, and international in its solidarities. Finally, and in a different vein altogether, Elizabeth DeLoughrey, also featured in this issue, draws on Caribbean writers such as Édouard Glissant and Kamau Brathwaite, among others, to figure the maritime regions of the Atlantic as "heavy waters," which she defines as "an ocean stasis that signals the dissolution of wasted lives."[10] Rather than posit the ocean depths as a void of no importance to human history, DeLoughrey demonstrates how, for descendants of Africans forcibly brought to the New World, the Atlantic does not connote enhanced freedoms and mobility of sail, as it did for early European novelists and explorers, but serves as a monument to the millions dead. The ocean describes a morass and limit to knowledge, but it is also an integral part of human history.[11] The Latin word *vastus*, as DeLoughrey points out, signifies not only the oceanic scale but also waste, and this doubled meaning extends into the present day as she traces the parallels between slave migrations and the contemporary treatment of refugees from all parts of the global South. The state plays a pivotal role in patrolling the seas and signaling the worth of these peoples. *Waste* also references the sea as "empty" and therefore the sense that its depths are of no concern to humans. This belief makes the sea available to absorb nuclear and other by-products of industrialization without fear of reprisal. But of course, as DeLoughrey argues, waste, whether human or otherwise, is precisely what constitutes the heavy, toxic waters of the Atlantic and beyond in a manner that endangers all planetary life.

In 2017 the American Comparative Literature Association forum "Oceanic Routes," edited by Isabel Hofmeyr—a contributor to this issue—and Kerry Bystrom, moved the conversation from a largely North Atlantic regional focus to understudied areas of the South Atlantic, Indian, and Pacific Oceans. In addition, they emphasize the material and geophysical dimensions of the sea that, as Blum says, allow for "a new epistemology—a new dimension—for thinking about surfaces, depths, and the extra-terrestrial dimensions of planetary resources and relations."[12] In addition, Hofmeyr and Bystrom address the issue of imperial power and dominance at sea, a focus underscored by their neologism *hydrocolonial*, which signals, as they put it, "an affinity with postcolonialism so crucial to the Southern regions we engage."[13] The legacy of imperialism persists in the ways that we imagine the seas geographically. In that same collection, Alice Te Punga Somerville reminds us that the Pacific Ocean was called El Mar del Sur (the South Sea) by its first European "discoverer," Vasco Núñez de Balboa, who then claimed all lands touching that ocean for Spain. Its status as a belated (Southern) addition to ocean studies, which began for academics, as mentioned above, in institutions in the West with the Mediterranean and Atlantic, is belied by Oceanic peoples' Indigenous history as having always been oriented to the sea and having always practiced an oceanic turn.[14] Taking heed of this Northern bias that renders Oceania a local particularity and afterthought, an afterword by Rachel Price suggests that "categories proper to aesthetics [might] help us grasp otherwise unrepresentable or unassimilable climate change and crisis, without allegorizing or merely miming a human/nonhuman relation. Such work offers one way that certain oceanic studies might continue to be specifically rooted in the oceans without their materiality becoming an imaginative constraint."[15] It is also precisely here that scholars of film, art, and literature, such as those gathered in this *ELN* issue, can provide their greatest provocations. An advantage of film, art, and literature as they mediate science, politics, history, geography, and maritime environments lies in their imaginative engagement in these discourses and materialities. The arts can imagine alternatives to the long history of the domination of nature. Price reminds us that the arts often distance us from reality in a manner that fosters alternative ways of thinking about human/nonhuman environments. The arts, as the contributors gathered in "Hydro-criticism" variously claim, imagine new feminist and queer relations between humans and nonhumans, create new kinds of aesthetics untethered to realist cartographies, and raise awareness of the imperial "hydro-power" that claims the seas for its own.

Hofmeyr begins our special issue with the South African Koleka Putuma's poem "Water" (2017). In her reading of Putuma's work, Hofmeyr reminds us that the leisure space of the seaside is underwritten by histories of dispossession, "launched from the littoral." Building on her previous work on "hydrocolonialism" mentioned above, Hofmeyr expands this term to mean not only colonization *by means of* water and colonization *of* water (the occupation of land with water resources, the appropriation of territorial waters, and the militarization of oceans) but also a colony *on* or *in* water (the ship as a floating penal colony and infamous penal islands such as Robben, Alcatraz, Rikers, Devil's, St. Helena, and Isabela). In addition, the term *hydrocolonial* names housing estates on hydroelectric plants in Canada and is a misspelling of *hydrocolonic*, both of which—the latter unwittingly—express the effects of hydrocolonialism as environmental degradation and colonial control, while *hydrocolonic* adds fears of disease and contamination from the flows of racialized peoples and goods through maritime travel and commerce. The term *hydrocolonial* underscores the uneven histories of the oceanic: how the seas might mean leisure, desire, and adventure, on the one hand, and fears of drowning, flooding, and contamination, on the other. To address the latter, Hofmeyr tracks the circulation of print (mail, newspapers, books, pamphlets, logbooks, journals, letters) as they are carried, written, and read on board ship and were policed in ports of call. Ports are vital checkpoints and stopgaps to the influx of seditious literature, revolutionary ideas, and people, feared by the state to be carriers of disease and contamination. The importance of gateways to national territories or, as Harris Feinsod relates, to other oceans via canals, reveals that as much as the seas have been lauded for enabling "flows," "circulation," and for serving as the connective tissue between geographies, nations, and languages, they are also regulated by means of ports, papers, inspections, quarantines, and refusals to prove insurmountable barriers. The seas remain racialized and colonized.

This political understanding of the seas is pressed further by DeLoughrey as she praises the recent turn to the seas as "wet matter," flows, fluidity, and mobility but cautions that "other, less poetic terms such as blue water navies, mobile offshore bases, high-seas exclusion zones, sea lanes of communication (SLOCs), and maritime 'choke points'" are perhaps more vital to understanding the geopolitics, hydrocolonialism, and environmental destruction of the seas. She opens with a warning to readers that the largest naval military exercise in history occurred in the fall of 2018, when this issue was under way. These exercises included all Pacific Rim nations allied with the United States, an alliance extended to India in May 2018 when the US Pacific Command was renamed the US Indo-Pacific Command. This inclusion enlarged the Pacific Rim region to the western coast of the Indian subcontinent and encompasses 100 million square miles of ocean surface. DeLoughrey directs "the focus of *hydro-criticism* toward *hydro-power,* defined as energy, force, militarism, and empire." Not only does military power ensure the transfer of oil and other energy across the planetary seas, but militaries are the largest consumers of fossil fuels. Further, their naval exercises result in massive marine animal deaths, often from sonic pollution. Given the huge ecological ramifications of petromilitaries, DeLoughrey asks, following Amitav Ghosh's influential work, why there are so few petrofiction novels. She argues that Ghosh is looking in the wrong place: the human-centric scale of the European novel tradition cannot encompass planetary, multiscaled intra-actions. She suggests instead that Indigenous, feminist, and postcolonial literature can make imaginatively perceptible the new scales and geographies of hydrocolonialism and petro-economies. Her example of this capability is poetry by Craig Santos Perez. We are grateful to him for granting us permission to reproduce elements from the cover of his book of poetry *[saina]* for this issue's cover. The montage neatly captures the range of "Hydro-criticism's" interventions, from Indigenous to military understandings of the sea.

This issue's first section, "Hydro-power: Charting the Global South," concludes with Meg Samuelson and Charne Lavery's essay "The Oceanic South." They begin by noting that the hemispheric South contains 20 percent more surface-area water than the North and continues to be dominated by extractive economies and settler colonialism. But, they continue, the Southern Hemisphere offers both centrality and prescience for considering maritime planetarity. Evoking the turbulent, three-dimensional space of southern seas, they argue for the importance of a regional geography that interlinks the Atlantic, Pacific, and Indian Oceans as well as the frozen continent of Antarctica. Drawing on fiction by J. M. Coetzee, Zakes Mda, and Witi Ihimaera, Samuelson and Lavery draw attention to how these fictions chart the mobility of the South Seas as well as the intersecting itineraries of interspecies migrations. Endangerment and mass slaughter, whether of penguins or of whales, refer us back to the long history of global resource extraction that constitutes the South, but literary evocations of their experience also express an alterity that is underived from humans and that, as Samuelson and Lavery argue, is alive to "more-than-human" oceanic materialities and multispecies interactions.

One reader who reviewed an article for this issue expressed the concern that when maritime literary and cultural scholars textualize the waters, they reduce the oceanic to a fashionable metaphor. Among these eminent scholars, this is decidedly not the case. First, as all the contributors assert—especially Cohen in her work on undersea film technologies—the alien environment of the sea requires that it be mediated through language, cameras, scientific instruments, and prostheses. In the next section, "Hydro-imaginaries," the essays draw our attention to the work of mediation: how what we know about the oceans is partial, situated, and shot through with imagination and desire. In her fascinating essay, Cohen unearths the film technologies and mise-en-scène by which filmmakers oriented audiences to the alien depths of the sea in the early postwar underwater films Disney's *Twenty Thousand Leagues under the Sea* (1954) and Jacques Cousteau and Louis Malle's *Silent World* (1956). These familiarizing techniques then allowed the filmmakers to highlight the spellbinding, otherworldly qualities of this realm, including an infamous shark feeding frenzy of a baby tooth whale that Cousteau's boat inadvertently killed with its propellers. While Cohen rightly criticizes the sensationalized portrayal of the sharks' natural behavior in the film (Cousteau's boat, after all, had killed the whale), her essay also raises an implicit question: What are other possible apparatuses (Indigenous, animal, microbial, feminist, queer, utopian) that mediate the seas and what alternatives do they make visible and/or knowable?

Often these alternatives are ignored at our peril, as Blum's essay shows. "Arctic Nation" proposes that "hydro-criticism" should take as its first principle "the elemental fluidity of the seas" such that "the undulating, nonhuman, nonplanar depths of the sea [serve] as a model for critical expansiveness." Rather than territorialize the seas, as hydrocolonialism and, on a lesser scale, underwater film techniques are wont to do, the oceanic "has a scalar fluidity that enables the hydrographic world to be at once global and microecological." Echoing, I believe, Samuelson and Lavery's claims for the planetary nature of the seas, Blum focuses the rest of her essay on a cautionary tale: the disastrous mission of the US-sponsored Lady Franklin Bay Arctic Expedition (1881–84) and the particular fate of Charles Buck Henry on Ellesmere Island, the northernmost island in the Canadian archipelago. Refusing to provision with local foods or engage in collective strategies of survival, most of the crew died, Henry was executed for stealing food, and many resorted to cannibalism. Blum offers this chilling tale as exemplary of "the flaws in importing a proprietary Western nationalism to an oceanic space."

Sometimes the grim administration of hydrocolonial power is best repulsed with a dose of humor. Teresa Shewry's essay on the darkly comic absurdity of Australian novelist Jane Rawson's futurist novel *A Wrong Turn at the Office of Unmade Lists* argues that humor arises when incongruity is pointed out. The novel is set in Melbourne, a city vulnerable to sea-level rise as well as to drought, heat waves, flooding, and fires. Its humor, Shewry argues, aims at "people's radically disproportionate contributions and exposures to the violence of sea-level rise and to precarious freshwater conditions." Her reading of the novel also tracks characters' strategies of survival, ways that unseen forces limit human intentions and apprehensions, and how ambivalent interdependences arise out of resource disparity. The novel's humor breaks the surface of uniformity—as a hydro-critical reading project must—revealing "intricate patterns of constraint, disproportion, and disruption."

Jeremy Chow and Brandi Bushman, two new voices in oceanic studies, take our hydro-critical reading project in a different, but highly necessary, direction: the question of embodiment and intraspecies relations. Referring us back to where we began this introduction, in the churning, mobile waters that intra-act with bodies, Chow and Bushman argue for a "hydro-eroticism" that "emphasize[s] erotic connectivities that bodies of water foster, while also remaining mindful of how fluid affiliations are policed, inhibited, or persecuted." They draw on queer theory and feminism to read Hulu's smash-hit television series *The Handmaid's Tale* and the

2017 Oscar award-winning film *The Shape of Water*. Their lively and critically incisive argument extends the analysis of "hydro-power" in the first section, and in Blum's essay, but moves away from large-scale planetary geopolitics to burrow down into questions of more-than-human embodiment. Water, in this essay, is an active participant in intimacies and pleasures. Chow and Bushman deploy Barad's concept of intra-activity to read water as it seeps through bodies and joins various participants and sites of sexuality in a manner that underscores human/nonhuman interdependencies and affirms queer pleasures against the ever-present backdrop of violent suppression.

What I was not entirely prepared for in editing this special issue was the prevalence of essays I received concerning state power, such that a major portion of "hydro-criticism" consists in anatomizing the workings and forms of "hydro-power." Even those essays that concern the seemingly more mundane genres of underwater adventure films and the comic novel indirectly invoke submarine warfare, Navy SEALs, and municipal decisions concerning equitable access to freshwater and fostering environmentally sustainable cities. Our historical moment has indeed shifted, which also occasions new parallels to the past. It was not only my own research interests in modernism that prompted the third section of this special issue, "Hydro-critical Practices: Modernism and the Sea," but the fact that in the recent challenges to US empire, one cannot but hear uncanny echoes of modernism, an era that witnessed the end of European hegemony. Questions of state power and empire have returned with a vengeance. Not only that, but I was curious to understand why sustained work on maritime modernism was, until recently, not widely embarked on. As one contributor, Nicole Rizzuto, affirms, the early twentieth century is "a dead zone in the history of seafaring literature." Against the romantic heyday of sail in the eighteenth and nineteenth centuries, the late nineteenth century saw the rise of coal-powered steamers, maritime workers' unions, and technologies such as radio and electronic navigation that lessened the wild freedoms (that is, the unregulated and horrendous violence) of sail at sea. In addition to the technological curtailments of what Carl Schmitt calls the "free sea," the extended imperial crisis of the modernist era multiplied existing controls and obstructions to maritime travel.[16] Feinsod, in his essay on the modernist era of the Panama Canal, neatly captures the range of obstructions of global flows and maritime cosmopolitanisms that have characterized the early to mid-portion of the twentieth century. These include "customs houses, immigration bureaus and passport control, deferred wages, state surveillance, and world war." Feinsod collocates the question of maritime circulation with the question of world literature and of how to write a comparative literary history. The proximity between an environment and a virtual community of letters makes sense, he says, only if we draw on the comparative terms Erich Auerbach lays out: "a tension among 'diverse backgrounds' converging on a 'common fate,' accessed by a method that seeks out multiple 'points of departure' coalescing around a synthetic or 'coadunatory' intuition." To this last gasp of European idealism, Feinsod, quoting Aamir Mufti, adds the postcolonial caveat that, akin to the harsh realities of maritime travel, literary critics have shown that world literature is "a regime of *enforced* mobility and therefore of *immobility* as well." Following the history of financial speculation—the booms and busts—of the Panama Canal Zone and weaving together literary accounts of this zone by Blaise Cendrars, Claude McKay, Eric Walrond, Wallace Stevens, and the Nicaraguan poet Ernesto Cardenal, Feinsod tracks the "conflicts, disparities, and experiments in sovereignty that best [characterize] the Panama Canal as a peculiar choke point of maritime globalization." This analysis refuses to equate all forms of sovereignty as Feinsod plumbs the depths of the "suction sea" with a reading of Walrond's short story "Wharf Rats" from *Tropic Death* (1926). There wealthy passengers on imperial liners toss mixed currency into shark-infested waters, where young boys dive for it among the wreckage of previous investment schemes. The Panama genealogy that Feinsod traces produces a "spatial form of capitalism" that "invite[s] us to reassemble a vision of worldwide modernism." While centered on one particular zone, this assemblage allows us to see the shards of unevenly distributed agencies across the globe: absent financial speculators in global centers, West Indian divers, Anglo-American tourist-spectators, modernist writers, and the material forces that shaped the canal.

Next, Rizzuto addresses the seeming dearth of seagoing fiction in the modernist era. After the subject of Conrad's fiction shifts to urban anarchists, Latin American silver mines, and revolutionaries in Switzerland and Russia, it is hard to think of his successors. In fact, Rizzuto writes, there are a number of forgotten or neglected maritime novelists who focus on working-class mariners, failed itineraries, and disorienting stasis. She situates these narratives amid "the restructuring of global circuits of power and attendant transformations of ways of seeing by presenting the waters as cramped with networks that blindside travelers and stymie movement." Focusing here on the British author James Hanley's *Hollow Sea*, which looks back from 1938 upon the First World War, Rizzuto elaborates on how Hanley references Conrad but alters his topics to account for merchant sailors and ships that were conscripted into military operations. The novel details how a merchant ship was ordered to transport fifteen hundred troops to the Dardanelles in a failed wartime operation. The sea, Rizzuto writes, is rendered opaque and deadly, not least because navigation was centralized by wartime control—"network-centric warfare"—that turns out to have been neither clear nor complete. In sum, the very technological means of navigation and communication that once allowed sailing on the open seas "have become imprisoning, confining, and disorienting" in the context of modern industrialized naval warfare. Even the seas themselves—sabotaged by U-boats, infiltrated by explosives, blocked by competing sea powers—are crowded and dangerous.

Maxwell Uphaus's essay concludes this cluster topic and allows us to circle back, in hydrospheric fashion, to where we began, with the early development of academic hydro-criticism itself. Noting the prevalence of epigraphs and references to T. S. Eliot's "The Dry Salvages" in work by C. L. R. James, Peter Linebaugh, and DeLoughrey, he pursues the strange conundrum posed by the attraction to the poetry of the conservative, insular Eliot on the part of progressive, even radical, Caribbean scholars.[17] One obvious reason is that "The Dry Salvages" highlights the sea's role in transcending national frames and establishing vast spatiotemporal scales that outstrip an androcentric focus. Most important, the poem's representations of the sea are shot through with many histories: nation building, empire founding, and their subsequent eclipse. The sea, in Uphaus's words, "simultaneously make[s], memorializ[es], and eclips[es] human histories." The sea, thus, is both constitutive of British history and the force that undoes it, a force that includes the history of the slave trade. Eliot's poem displays a defensive posturing that, in portraying Britons as "victims of their own seaborne expansion," not only suppresses but supplants the dead of the Middle Passage. Moving from the Atlantic to the Mississippi and its legacy in the slave trade, to India, Eliot's poem represents different facets of the same imperial oceanic history. And while he minimizes the violence of the British in this history, Uphaus argues that the poem cannot keep this violence completely suppressed. It motivates how Eliot "reenvision[s] oceanic history as an ongoing, catastrophic process that leaves an ocean strewn with wastage." The poem, in this reading, questions both seaborne imperialism and insular nationalisms and thus can well serve as a springboard for later counterhistories of the seas.

This issue concludes with a review essay by Allison Nowak Shelton, "Learning from Rivers: Toward a Relational View of the Anthropocene." Shelton reminds us that rivers are an integral part of the planetary hydrosphere as well as an important bridge between deep geological time and shorter-term "surface" history, scientific data, and qualitative affective experiences, including myth and storytelling. They are also frequently the center of heated political and environmental struggles as well as studies that cross many academic disciplines. After reviewing a monograph and two important recent essay collections, Shelton ends with a reading of the Meenachil River in Arundhati Roy's *The God of Small Things* (1997) and argues that cultural, political, and economic conceptions of rivers "are threads of a complex environmental network that determines and is determined by river ecology." Taken together, the wide-ranging essays that "Hydro-criticism" comprises demonstrate the critical advantages learned from the scalar fluidity and ineluctable materiality of water. Gleaning new perspectives, ontologies, and transmaterial subjectivities from the vantage point of the seas, these essays transform critical paradigms, artistic and reading practices, and erotic pleasures as well as draw our attention to maritime ethical and political urgencies.

LAURA WINKIEL is associate professor in the English Department at the University of Colorado, Boulder. She is author of *Modernism: The Basics* (2017) and *Modernism, Race and Manifestos* (2008), and is coeditor of *Geomodernisms: Race, Modernism, Modernity* (2005). She is presently at work on a book tentatively called *Compost Fictions: Vernacular Practices and the Shifting Ground of the Human*. She was president of the Modernist Studies Association in 2017–18.

Notes

1. A few of the most influential Indigenous and global South nonfiction writers on the sea include Hau'ofa, Brathwaite, and Glissant (Somerville, *Once*). See also DeLoughrey, *Routes*; and Shewry, *Hope*.
2. Raban, *Oxford*, 28–29.
3. Helmreich, *Alien Ocean*, 17.
4. The term *intra-action* is used by Barad to underscore how subjects and objects are not preestablished entities but are derived from an interrelated "dynamism of forces" (*Meeting*, 141).

Intra-action affirms the impossibility of separately derived and existing entities. These terms, *material actant* and *posthumanism*, can be understood to reflect the scientific perception of how human actions and marine environments mutually impact and even coconstitute one another. Seas and humans intra-act such that humans are permanently entangled with their environments, marine and otherwise.

5 Alaimo writes: "Trans-corporeality situates the (post)human as always already part of the world's intra-active agencies. For an oceanic sense of trans-corporeality to be an ethical mode of being, the material self must not be a finished, self-contained product of evolutionary genealogies but a site where the knowledges and practices of embodiment are undertaken as part of the world's becoming" (*Exposed*, 127).

6 See Nixon, *Slow*.

7 Yaeger, "Editor's Column," 530.

8 Cohen, "Literary Studies," 658; see also Sekula, *Fish*.

9 Blum, "Prospect," 671.

10 DeLoughrey, "Heavy," 703.

11 DeLoughrey, "Heavy," 704.

12 Blum, "Introduction," 151. See also Steinberg and Peters, "Wet."

13 Bystrom and Hofmeyr, "Oceanic Routes," 3.

14 Somerville, "Where Oceans," 28.

15 Price, "Afterword," 51–52.

16 Schmitt, *Nomos*, 98–99.

17 James, *Black*; Linebaugh, "All"; DeLoughrey, "Heavy."

Works Cited

Alaimo, Stacy. *Exposed: Environmental Politics and Pleasures in Posthuman Times*. Minneapolis: University of Minnesota Press, 2016.

Barad, Karen. *Meeting the Universe Halfway: Quantum Physics and the Entanglement of Matter and Meaning*. Durham, NC: Duke University Press, 2007.

Blum, Hester. "Introduction: Oceanic Studies." *Atlantic Studies* 10, no. 2 (2013): 151–55.

Blum, Hester. "The Prospect of Oceanic Studies." *PMLA* 125, no. 3 (2010): 670–77.

Brathwaite, Edward Kamau. *History of the Voice: The Development of Nation Language in Anglophone Caribbean Poetry*. London: New Beacon, 1984.

Bystrom, Kerry, and Isabel Hofmeyr. "Oceanic Routes: (Post-It) Notes on Hydro-colonialism." *Comparative Literature* 69, no. 1 (2017): 1–6.

Cohen, Margaret. "Literary Studies on the Terraqueous Globe." *PMLA* 125, no. 3 (2010): 657–62.

DeLoughrey, Elizabeth. "Heavy Waters: Waste and Atlantic Modernity." *PMLA* 125, no. 3 (2010): 703–12.

DeLoughrey, Elizabeth. *Routes and Roots: Navigating Caribbean and Pacific Island Literature*. Honolulu: University of Hawai'i Press, 2007.

Glissant, Édouard. *Poetics of Relation*. Ann Arbor: University of Michigan Press, 1997.

Hau'ofa, Epeli. *We Are the Ocean: Selected Works*. Honolulu: University of Hawai'i Press, 2008.

Helmreich, Stefan. *Alien Ocean: Anthropological Voyages in Microbial Seas*. Berkeley: University of California Press, 2009.

James, C. L. R. *The Black Jacobins: Toussaint L'Ouverture and the San Domingo Revolution*. 2nd ed. New York: Vintage, 1989.

Linebaugh, Peter. "All the Atlantic Mountains Shook." *Labour/Le Travail*, no. 10 (1982): 87–121.

Nixon, Rob. *Slow Violence and the Environmentalism of the Poor*. Cambridge, MA: Harvard University Press, 2011.

Price, Rachel. "Afterword: The Last Universal Commons." *Comparative Literature* 69, no. 1 (2017): 45–53.

Raban, Jonathan. "Introduction." In *The Oxford Book of the Sea*, edited by Jonathan Raban, 1–34. New York: Oxford University Press, 1992.

Schmitt, Carl. *The* Nomos *of the Earth in the International Law of the* Jus Publicum Europaeum. New York: Telos, 2006.

Sekula, Allan. *Fish Story*. Düsseldorf: Richter, 2003.

Shewry, Teresa. *Hope at Sea: Possible Ecologies in Oceanic Literature*. Minneapolis: University of Minnesota Press, 2015.

Somerville, Alice Te Punga. *Once Were Pacific: Māori Connections to Oceania*. Minneapolis: University of Minnesota Press, 2012.

Somerville, Alice Te Punga. "Where Oceans Come From." *Comparative Literature* 69, no. 1 (2017): 25–31.

Steinberg, Philip, and Kimberley Peters. "Wet Ontologies, Fluid Spaces: Giving Depth to Volume through Oceanic Thinking." *Environment and Planning D: Society and Space* 33, no. 2 (2015): 247–64.

Yaeger, Patricia. "Editor's Column: Sea Trash, Dark Pools, and the Tragedy of the Commons." *PMLA* 125, no. 3 (2010): 523–45.

Provisional Notes on Hydrocolonialism

ISABEL HOFMEYR

Abstract This article discusses the meanings and applications of the rubric *hydrocolonialism*. Starting with a South African poem as an example, the piece sets out a definition of the term before outlining the existing literary scholarship that could fall under its umbrella. The article then turns to discuss hydrocolonial book history and what we might learn by tracing books on their oceanic journeys. One important node in these journeys was the port city, where customs and excise officials examined texts to see whether they were pirated, seditious, or obscene. These inspectors in effect functioned as censors and adjudicators of copyright. The article examines how these protocols worked in practice and concludes by discussing this dockside mode of reading as a hydrocolonial literary formation.
Keywords hydrocolonialism, Koleka Putuma, oceanic book history, colonial copyright, customs and excise

Koleka Putuma is a South African performance artist, poet, and theater director. One of her best-known performance pieces, "Water," recently made its way into her debut poetry volume, *Collective Amnesia*.[1] The poem refracts a history of South Africa through the seashore. It opens with memories of rare and anxious visits to the beach, with fretting adults forbidding children from going in too deep, "as if the ocean had food poisoning." The speaker recalls "this joke / about Black people not being able to swim, / or being scared of water" and wonders why, every time she sees the sea, she feels as if she is drowning.

After this opening, set in the present, the poem swivels to the past, outlining a repressed history of slavery and colonization that returns via the sea itself:

> every time our skin goes under,
> it's as if the reeds remember that they were once chains,
> and the water, restless, wishes it could spew all of the slaves
>   and ships onto shore,
> whole as they had boarded, sailed and sunk.
> Their tears are what has turned the ocean salty.

57:1, April 2019 DOI 10.1215/00138282-7309644

The remainder of the poem toggles between past and present, providing a series of histories configured from the shoreline. The present-day beach constitutes a zone of whiteness and consumerism: "For you, the ocean is for surfboards boats and tans / and all the cool stuff you do under there in your bathing suits and goggles." At the same time, Putuma reminds us that this practice of leisure is underwritten by earlier histories of dispossession launched from the littoral, the bridgehead of maritime imperialism and land invasion.

Cognizant of such histories, a new generation of black beachgoers shares a more somber sense of purpose. No longer plagued by anxiety, this generation visits the beach conceptualized now as a site of pilgrimage:

> We have come to be baptized here.
> We have come to stir the other world here.
> We have come to cleanse ourselves here.
> Our respect for water is what you have termed fear.
> The audacity to trade and murder us over water
> Then mock us for being scared of it.

The poem concludes in the time of postapartheid South Africa and its galling orthodoxies of reconciliation that require black South Africans "to dine with the oppressors / and serve them forgiveness," after which everyone can "wash this bitter meal with amnesia." The final lines read: "And go for a swim after that. / Just for fun. / Just for fun."

Putuma herself was born in 1993, on the eve of South Africa's political transition to democracy, and forms part of a generation of "born frees" who have had to take on the task of confronting the omnipresent aftermaths of apartheid and racialized inequality. This generation also led the recent student movement known as Rhodes Must Fall (in Cape Town) and Fees Must Fall (in Johannesburg), which included demands for the pervasive structures of white privilege to be dismantled. In this climate, the heroic antiapartheid narrative that posited the rainbow nation as its end point has grown brittle.

One feature of this nationalist antiapartheid discourse was a turning away from the ocean, the site of imperial incursion, toward the land, the locus of the desired nation. More recently, a group of postapartheid black feminist scholars (of which Putuma forms a part) has sought to shift this balance, by reclaiming the ocean from a decolonizing perspective, foregrounding the histories of slavery that it brought, but also examining how one might unseat and reimagine these genealogies.[2] The Indian Ocean in particular has become an important postapartheid matrix that has allowed scholars to explore archives of connections that exceed the nation-state through tracing histories of slavery and indenture.[3]

Putuma's poem can usefully be read as part of this oceanic turn, her text an attempt to decolonize the ocean, refracting the different levels of colonial control exerted by means of, over, and through water while reflecting on how black communities and ancestors have dealt with the ocean. An apposite term for this endeavor might be *hydrocolonialism*, a neologism whose meanings I have set out elsewhere.[4]

These meanings could include (1) colonization *by way of* water (various forms of maritime imperialism), (2) colonization *of* water (occupation of land with water resources, the declaration of territorial waters, the militarization and geopoliticization of oceans), and (3) a colony *on* (or in) water (the ship as a miniature colony or a penal island). While the word *hydrocolonialism* is an invention, there are two related uses that I have encountered on the web. The first is *hydrocolony*, a Canadian term for a housing estate for workers on a hydroelectric plant.[5] The second is a grammatically incorrect synonym for hydrocolonic, that is, colonic irrigation that at times appears as hydrocolonial irrigation. Both of these raise pertinent themes: the first points to the "fundamental connection between water, its management, and the colonial or neocolonial relations in the modern era," as Sarah Pritchard argues in her account of hydroimperialism.[6] Designating the workers' housing estate as a colony speaks of an imperial imaginary in the management of water and the labor associated with it. Hydroelectric dams are showpieces of modernity, displacing communities and affecting aquatic ecologies and river flows. The hydrocolony consequently speaks to themes of colonial control and environmental degradation. The term *hydrocolonial/hydrocolonic irrigation* resonates with these themes by suggesting accelerated processes of waste making. However erroneous the term, it does capture metaphorically the waste-making systems of colonial rule, where certain people were rendered as waste, whether through the slave trade, indenture, or penal transportation. Imperial border regimes fed off these imaginaries of waste as contamination, constituting themselves as quarantine ramparts against disease and "undesirable aliens." As this article argues, such hydrocolonial formations had far-reaching effects not only in terms of sustaining racialized ideologies but also in terms of how books and printed matter that had to pass through port-city regimes came to be defined.

Across these definitions, the term signals an affinity with postcolonialism while declaring an intention to shift the intellectual center of gravity away from a purely land-focused one. Just as the term *postcolonial* aims to understand a world shaped by European empires and their aftermaths, so *hydrocolonialism* signals a commitment to understanding a world indelibly shaped by imperial uses of water.[7] As global warming and climate change take hold, and as sea levels rise, we are reminded that we all live in the aftermath of the hydropolitics of imperialism. Another objective of postcolonial theory is to unmask imperial modes of power. So too, hydrocolonialism makes visible relations of power that have been shaped around water and its colonial appropriations, as Putuma's poem aptly demonstrates. Allied to this objective, postcolonial theory has always sought to decolonize knowledge by exposing the colonial contours that shape existing curricula and canons. Likewise hydrocolonialism seeks to lay bare colonial constructions and representations of water and to undo these. The title of the article has a further postcolonial entailment, since it invokes Achille Mbembe's "Provisional Notes on the Postcolony" (the earliest English iteration of what became *On the Postcolony*). This work stressed the complexities of a postslavery and postcolonial world and what he termed the chaotic plurality of the postcolony.[8] The term *hydrocolonialism* could be construed in a cognate way, namely, that we live with the aftermaths of post-hydro-imperialism, which has produced and will continue to produce a chaotic plurality of ecological disorders.

...

What work might this term *hydrocolonialism* do and how might it be deployed?

There is of course a growing body of literary scholarship (of which this volume provides one example) that addresses hydrocolonial concerns. Literary scholars have begun pioneering a range of methods and ways of reading to comprehend the oceans and water more generally from a nonhuman perspective. This work has produced a rich array of ideas, like amphibious aesthetics, littoral form, monsoon assemblages, heavy waters, hydropoetics, submarine aesthetics, transcorporeality, and sea ontologies—all concepts that push us closer to a material engagement with water.[9] Caribbean theorists have long furnished us with a tradition of thinking about the imperial ocean hydropoetically. Whether through notions of tidalectics, the haunted ocean, or the sovereignty of the drowned, these thinkers have offered a rich range of ideas for thinking with and through water. Elizabeth DeLoughrey discusses the "heavy waters of [Atlantic] ocean modernity" and the waste that they produce both in the form of drowned slave lives and in the current militarized pollution of the Atlantic, on whose seabed rest several nuclear reactors and warships.[10] More recently, scholars of the Indian Ocean have started to explore more material approaches: Lindsay Bremner's idea of "monsoon assemblages" investigates the interlinked environmental, oceanic, and infrastructural histories of Indian Ocean cities; Charne Lavery's work analyzes how the depths of the Indian Ocean have been encountered, imagined, and represented.[11] Much ocean-related literary criticism manifests a deepening engagement with the materiality of water, going below the waterline to explore historical and aesthetic submarine themes, be these the underwater sculptures of Jason deCaires Taylor or the history of sharks and slave ships.[12]

One way in which to extend these literary critical frameworks is to think about both the book and the ocean in material idioms. If we trace the hydrocolonial passage of books, would this make visible new forms of reading? In part, this question has been answered by the considerable body of scholarship on the ship as a configuration of reading and writing. Ships were documentary machines that generated logbooks, sailors' journals, and passengers' handwritten newspapers.[13] Royal navy vessels were "floating secretariats" undertaking scientific work and marine exploration.[14] Books traveled as cargo, in ship's libraries, with sailors and passengers.[15] Whether Margaret Cohen's analysis of shipboard writing and its trajectories into the adventure novel, or Hester Blum's account of what and how sailors in the Atlantic read on board, such scholarship has directed our attention to the ship as a rich textual space.

Shipping tempos and routes shaped the forms of colonial literature. Each mail ship delivered bundles of periodicals, which were in turn cut and pasted into local publications, so that the latter swelled in the wake of ships' visits and shrank in their absence. The lulls between the coming and going of the ocean mail ships acted as one prompt for the production of colonial writing.[16]

As volumes traveled by sea, their material form could be changed. If packed in the hold, books could be affected by other cargo: the smell of oil penetrated paper as did that of areca nut, while salt buckled pages. Books that were badly packed could be battered and scarred in transit. Steamships required coal and moved between bunkering ports, where the vessel would be fueled, at times by slave or forced labor.[17] Coal dust clouded the air and penetrated every nook and cranny, so that any books on board would in all likelihood have become gritty. To receive a book with the marks of its journeys was to be reminded of the routes it had traveled and possibly the labor that had enabled its passage (although this insight was probably limited to those who had knowledge of coaling practices or had relations or acquaintances who had done such work).

Yet the hydrocolonial journeys of books extended beyond the ship itself, especially in the port city, a choke point through which printed matter from outside a colony had to pass. Within the port itself, customs examiners checked printed material to see that it was not pirated, seditious, or obscene, a procedure that turned officials into censors as well as adjudicators of copyright policy. Port cities constituted a hydroborder, which had important implications for how books came to be defined, regarded, and classified.[18] The study of the maritime boundary-making of colonial states in the British Empire has shown the border's seminal function in terms of creating racialized identities, paranoid styles of governmentality, and epidemiological forms of statecraft.[19] This scholarship has mainly attended to the practices of immigration restriction, a phenomenon of the 1880s on. A much older branch of port-city governance was customs (subsequently melded with excise), an institution with powerful implications for books and reading practices.

These textual functions of customs and excise were determined in part by law and regulation but as much by the daily protocols of the dockside. The law on copyright comprised international, imperial, and colonial legislation and was contradictory and confused. It was difficult to decide which law applied where, and officials instead made ad hoc decisions governed by the logic of everyday procedure. Hence they tended to treat copyright less as an abstract legal entitlement than as a sign of manufacture or origin. Mark of origin ("Made in England," "Made in Australia," etc.) was an important aspect of late nineteenth-century British imperial trade, and commodities were required by the empire-wide Merchandise Marks Act of 1887 to carry such marks. Customs officials consequently paid considerable attention to them when checking commodities.[20]

In dealing with books and questions of copyright, customs relied on the logic of "marks of origin" to make decisions. Unable to fathom the various levels of copyright law, officials sought evidence that the book had been "composed, manufactured or copyrighted" in Britain, subjecting the book to a logic of origin and source. In these procedures, they were supported by the Merchandise Marks Act, which specifically indicated that British copyright could "be taken to be [an] indirect indication of British manufacture."[21] Their obsession with marks of origin resonated with immigration-restriction procedures of excluding those with the wrong bodily "marks of origin" and the notorious writing and dictation test by means of which would-be immigrants, as a condition of entry, could be required to write a dictated passage in a European language and in Roman script.

Customs officials acquired powers of censorship, since they could seize or detain material deemed seditious or obscene. Like all censors, they read in a paranoid and suspicious way, a tendency increased by the uncertainty of the port city itself. Ports aimed to pave the ocean and assert sovereignty over the conjunction of land and sea. While any form of sovereignty is potentially flimsy, hydrocolonial modes are especially so, since they are subject to the ocean—both to the sea's physical laws and to the people, objects, and animals delivered by vessels docking in the port. These objects (and people) were always under suspicion: they might be diseased, contaminated, seditious, obscene, illegal, or counterfeit. This anxiety was underscored ecologically. Titles of various offices attached to the harbor—Port Captain, Water Police, Beach Magistrate, Beach Master, Receiver of Wrecks (the latter three governing shipwrecks, flotsam, and jetsam)—exuded authority even as they reminded their holders that they were dependent on the ocean and its physical vagaries, a danger to which custom houses, generally built on reclaimed land, were occasionally heir.

Books and other objects stopped for examination were regarded as though they carried the contamination of the ship with them and were at times treated as if they had dangerous microbial properties. Banned films were described as unfit for human consumption as though they carried bacterial infection; condoms with objectionable slogans on them were deemed "harmful to health"; indecent items were considered injurious to the public well-being, while undesirable publications apprehended in the port were likened to foreign bodies.[22] In Australia suspect magazines were regarded as physical contaminants, immune even to the bleaching agent used in the manufacture of their pages. As one official noted, this disinfectant should have been able to "militate against disease" but in the face of such obscenity was useless.[23]

When assaying print objects, one might anticipate that examiners paid most attention to the words in the publication under scrutiny. Yet in many instances writing was not necessarily prioritized, since it constituted only one dimension of the object. Instead, the printed object was apprehended in its entirety or adjudged by a range of material features. French novels were often categorized as undesirable simply for being French or because of their illustrations. Book covers provided another avenue for assaying a publication; an offending jacket sufficed to have the object banned or burned. In other instances officials followed a sampling method in which random passages from suspect texts were selected, rather like an excise man testing a consignment of alcohol.[24] As already indicated, officials apprehended objects in material ways, often imputing microbial properties to inorganic substances.

The reading protocols that these officials formulated and refined would be adopted and extended by subsequent censorship regimes in South Africa.[25] Growing anticommunism from the 1920s on gave customs further occasion to extend their scope. However, as the Cold War gained ground and became an international "security issue" and as more films circulated (something for which customs lacked viewing facilities), censorship was taken over by the Department of the Interior (with their specially built "censorship theatre") and then by a full-fledged censorship apparatus under the apartheid regime.[26]

Apartheid censorship has been widely studied and is quite correctly understood as a key plank in the regime's attempt to control dissent. At the height of this system, Nadine Gordimer observed that the South African censorship machine treated literature "as a commodity to be boiled down to its components and measured like a bar of soap at the Bureau of Standards."[27] Gordimer's analogy, intended to belittle the censorship board, compares it to the body in charge of quality control and standards, which had its origins in customs and excise. While Gordimer was probably unaware of the genealogies of censorship in South Africa, her comments nonetheless usefully summarize the origins of censorship, a process having its roots in the hydrocolonial practices of customs officials in port cities.

Placing books in a hydrocolonial framework makes visible the meanings and associations that they accrued as they journeyed over land and sea. Such meanings could reside in the pages of the book itself, smelling of areca nut or buckled by salt. Strong emotion could also cohere around an absent book—the publication or periodical that never arrived, having been detained at customs and possibly burned or destroyed there. Such missing books generated more powerful feelings than the volumes that did actually arrive.

These censorship protocols remind us that books were a form of soft imperial power, instruments of civilization, and calling cards of Englishness. They had consequently to be policed lest the dangerous ones infected their readers. Such suspect texts were "lively," humming with dubious elements rather like adulterated cargo. We might usefully describe the ways in which customs officials read as apprehensive. Not only did they apprehend the object by seeing it physically and mentally, but at times they seized and detained it on the grounds of being piracy, sedition, or obscenity. These objects elicited suspicion, producing fear or apprehensiveness for what they might contain.

...

The colonial condition is always nervous, both for the "native" and the settler. The nervous condition of the "native" has been much discussed, most notably in Tstitsi Dangarembga's Zimbabwean novel *Nervous Conditions*.[28] The condition of the boundary-making settler is equally nervous, or as this article has suggested, apprehensive. This condition was particularly acute on the hydroborder, where the "normal" anxieties of the boundary were exacerbated by ecological uncertainty, health hazards of ships arriving in port, and paranoia about "undesirable aliens" arriving by sea. The textual practices that these officials evolved in the uncertainty of the port city might be regarded as reading practices with hydrocolonial origins. These apprehensive reading modes were to be generalized into South African censorship practices, but their initial formation was tied up with maritime boundary-making.

Putuma's poem works in and against this genealogy. The poem opens with the nervousness that the white-controlled beach produces for an older generation. Putuma addresses this nervous condition not by rejecting the sea and embracing the land more vehemently. Instead, her poem recalibrates the ocean, suggesting ways in which it might be decolonized, by naming its hydrocolonial formation and invoking precolonial understandings of the sea and the littoral as an ancestral realm and site of pilgrimage, healing, and renewal. From the vantage of the littoral, Putuma invites us to consider southern Africa as a hydrocolonial formation. Along with the poem, this article has started to outline some of the new analytic possibilities that such a genealogy suggests.

ISABEL HOFMEYR is professor of African literature at the University of the Witwatersrand and Global Distinguished Professor at New York University. She has worked extensively on the Indian Ocean world and oceanic themes more generally. Recent publications include *Gandhi's Printing Press: Experiments in Slow Reading* (2013) and a special issue of *Comparative Literature* (2016) titled "Oceanic Routes," coedited with Kerry Bystrom. She heads up a project called Oceanic Humanities for the Global South with partners from Mozambique, Mauritius, India, Jamaica, and Barbados (www.oceanichumanities.com).

Acknowledgments

I acknowledge the support of the National Research Foundation, which enabled research undertaken toward this article.

Notes

1 Putuma, *Collective Amnesia*.
2 Gqola, *What Is Slavery to Me?*; Baderoon, "African Oceans."
3 Hofmeyr, "South Africa's Indian Ocean."
4 Bystrom and Hofmeyr, "Oceanic Routes."
5 "Hydro Colony" and "Hydro Colony—Ear Falls (1957)."
6 Pritchard, "From Hydroimperialism," abstract.
7 For apposite examples of this point, see Deckard, "Latin America"; and Pritchard, "From Hydroimperialism."
8 Mbembe, "Provisional Notes"; Mbembe, *On the Postcolony*.
9 These concepts are drawn from Samuelson, "Coastal Form"; Bremner, "Monsoon Assemblages"; DeLoughrey, "Heavy Waters"; DeLoughrey, "Submarine Futures"; Sandhu, "Hydropoetics"; Alaimo, *Bodily Natures*; and Cohen, *The Novel and the Sea*.
10 DeLoughrey, "Heavy Waters."
11 Bremner, "Monsoon Assemblages"; Lavery, "Indian Ocean Depths."
12 DeLoughrey, "Submarine Futures"; Rediker, "History from below the Waterline."
13 Delmas, "From Travelling to History"; Sheikh, "The Alfred and the Open Sea."
14 Killingray, "Introduction," 5.
15 Blum, *The View*; Maynard, "'In the Interests'"; Hyslop, "Steamship Empire."
16 Da Silva, "'Homeward Bound.'"
17 Anim-Addo, "'A Wretched.'"
18 I am grateful for a reader of the article who suggested the term *hydroborder*.
19 Dhupelia-Mesthrie, "False Fathers"; Heath, *Purifying Empire*; Hyslop, "Undesirable"; Klaaren, "Migrating"; Lake and Reynolds, *Drawing*; MacDonald, "Strangers"; MacDonald, "Identity Thieves"; Martens, "Pioneering."
20 Hofmeyr, "Colonial Copyright."
21 Payn, *The Merchandise Marks Act*, 21; Treasury Department 972, 936, "Seizure under the Copyright Protection and Books Registration Act of 1895 of Certain Books," 1906, Cape Town Archives Repository, Western Cape Provincial Archives and Records Services, Cape Town.
22 Customs and Excise 199, A/10/5X, "Prohibited and Restricted Imports. Indecent and Objectionable Articles," 1939, Central Archives Repository, National Archives Repository, Pretoria; Customs and Excise 199, A10/6X, "Indecent and Objectionable Articles," 1922, Central Archives Repository.
23 Heath, *Purifying Empire*, 115.
24 Lieutenant Governor 19, 25/54, "Customs Detention of Certain Books," 1906, Transvaal Archives Repository, National Archives Repository; Customs and Excise 199, A10/6X; Treasury Department 912, 2145, "Detention of Book *Vechten en Vluchten van Beyers en Kemp*," 1905, Cape Town Archives Repository.
25 Treasury Department 815, 1505, "Complaint by Mr. Speelman Regarding the Detention of Certain Books by the Customs," 1904–5, Cape Town Archives Repository; Treasury Department 912; Attorney General 1367, 296, "Detention of Book 'The Mobile Boer,'" 1904, Cape Town Archives Repository; Attorney

General 1441, 4790, "Book Entitled 'De Dochter van den Handsuffer [Handsopper]': Detention of," 1904, Cape Town Archives Repository.

26 Customs and Excise 209, A10/26X, "Prohibited and Restricted Imports. Censorship of Films. Precedent," 1917–57, Central Archives Repository; Secretary of the Treasurer 696, F4/71, "Department of Interior. Censorship: 1) Entertainments (Censorship) Act 28/1931 and Amendments. 2) Board of Censors: Appointment and Remuneration of Members. 3) Board of Censors: Staff for," 1963, Central Archives Repository.

27 University of the Witwatersrand, Historical Papers, Nadine Gordimer Collection, A 3367, F3, "Censorship in South Africa, Letter to the Secretary of the Interior," January 23, 1973.

28 Dangarembga, *Nervous Conditions*.

Works Cited

Alaimo, Stacey. *Bodily Natures: Science, Environment, and the Material Self.* Bloomington: Indiana University Press, 2010.

Anim-Addo, Anyaa. "'A Wretched and Slave-Like Mode of Labor': Slavery, Emancipation, and the Royal Mail Steam Packet Company's Coaling Stations." *Historical Geography* 39 (2011): 65–84.

Baderoon, Gabeba. "The African Oceans: Tracing the Sea as Memory of Slavery in South African Literature and Culture." *Research in African Literatures* 40, no. 4 (2009): 89–107.

Blum, Hester. *The View from the Mast-Head: Maritime Imagination and Antebellum American Sea Narratives.* Chapel Hill: University of North Carolina Press, 2008.

Bremner, Lindsay. "Monsoon Assemblages Research Project." November 21, 2015. www.westminster.ac.uk/news/2015/monsoon-assemblages-research-project.

Bystrom, Kerry, and Isabel Hofmeyr. "Oceanic Routes: (Post-It) Notes on Hydro-colonialism." *Comparative Literature* 69, no. 1 (2017): 1–6.

Cohen, Margaret. *The Novel and the Sea.* Princeton, NJ: Princeton University Press, 2010.

Dangarembga, Tsitsi. *Nervous Conditions.* London: Women's, 1988.

Da Silva, Sarah. "'Homeward Bound': Periodicity, the Cape Colony's Literary Culture, and the *Cape Monthly Magazine*." *English Studies in Africa* 57, no. 1 (2014): 9–20.

Deckard, Sharae. "Latin America in the World-Ecology: Origins and Crisis." In *Ecological Crisis and Cultural Representation in Latin America: Ecocritical Perspectives on Art, Film, and Literature,* edited by Mark Anderson and Zelia M. Bora, 3–19. Lanham, MD: Lexington, 2016.

Delmas, Adrien. "From Travelling to History: An Outline of the VOC Writing System during the Seventeenth Century." In *Written Culture in a Colonial Context: Africa and the Americas, 1500–1900,* edited by Adrien Delmas and Nigel Penn, 99–126. Leiden: Brill, 2012.

DeLoughrey, Elizabeth. "Heavy Waters: Waste and Atlantic Modernity." *PMLA* 125, no. 3 (2010): 703–12.

DeLoughrey, Elizabeth. "Submarine Futures of the Anthropocene." *Comparative Literature* 69, no. 1 (2017): 32–44.

Dhupelia-Mesthrie, Uma. "False Fathers and False Sons: Immigration Officials in Cape Town, Documents, and Verifying Minor Sons from India in the First Half of the Twentieth Century." *Kronos* 40, no. 1 (2014): 99–132.

Gqola, Pumla. *What Is Slavery to Me? Postcolonial/Slave Memory in Post-apartheid South Africa.* Johannesburg: Witwatersrand University Press, 2010.

Heath, Deana. *Purifying Empire: Obscenity and the Politics of Moral Regulation in Britain, India, and Australia.* Cambridge: Cambridge University Press, 2010.

Hofmeyr, Isabel. "Colonial Copyright, Customs, and Port Cities: Material Histories and Intellectual Property Comparative Literature." *Comparative Literature* 70, no. 3 (2018): 264–77.

Hofmeyr, Isabel. "South Africa's Indian Ocean—Notes from Johannesburg." *History Compass* 11, no. 7 (2013): 508–12.

Hydro Colony. wikimapia.org/25338388/Hydro-Colony (accessed August 18, 2018).

Hydro Colony—Ear Falls (1957): The Gateway to Northwestern Ontario History. Ear Falls Public Library. images.ourontario.ca/gateway/55734/data (accessed August 19, 2018).

Hyslop, Jonathan. "Steamship Empire: Asian, African, and British Sailors in the Merchant Marine c. 1880–1945." *Journal of Asian and African Studies* 44, no. 1 (2009): 49–67.

Hyslop, Jonathan. "'Undesirable Inhabitant of the Union . . . Supplying Liquor to Natives': D. F. Malan and the Deportation of South Africa's British and Irish Lumpen Proletarians, 1924–1933." *Kronos* 40, no. 1 (2014): 178–97.

Killingray, David. "Introduction: Imperial Seas: Cultural Exchange and Commerce in the British Empire, 1780–1900." In *Maritime Empires: British Imperial Maritime Trade in the Nineteenth Century,* edited by David Killingray, Margarette Lincoln, and Nigel Rigby, 1–12. Woodbridge: Boydell Press and National Maritime Museum, 2004.

Klaaren, Jonathan Eugene. "Migrating to Citizenship: Mobility, Law, and Nationality in

South Africa, 1897–1937." PhD diss., Yale University, 2004.
Lake, Marilyn, and Henry Reynolds. *Drawing the Global Colour Line: White Men's Countries and the International Challenge of Racial Equality.* Cambridge: Cambridge University Press, 2008.
Lavery, Charne. "Indian Ocean Depths: Cables, Cucumbers, Consortiums." Paper presented to the colloquium "Indian Ocean Energies," University of the Witwatersrand, Johannesburg, July 2016.
MacDonald, Andrew. "The Identity Thieves of the Indian Ocean: Forgery, Fraud, and the Origins of South African Immigration Control, 1890s–1920s." In *Registration and Recognition: Documenting the Person in World History*, edited by Keith Breckenridge and Simon Sretzer, 253–76. London: British Academy, 2014.
MacDonald, Andrew. "Strangers in a Strange Land: Undesirables and Border-Controls in Colonial Durban, 1897–c. 1910." MA thesis, University of KwaZulu-Natal, 2007.
Martens, Jeremy C. "Pioneering the Dictation Test? The Creation and Administration of Western Australia's Immigration Restriction Act, 1897–1901." *Studies in Western Australian History* 28 (2013): 47–67.
Maynard, John. "'In the Interests of Our People': The Influence of Garveyism on the Rise of Australian Aboriginal Political Activism." *Aboriginal History* 29 (2005): 1–22.
Mbembe, Achille. *On the Postcolony.* Berkeley: University of California Press, 2001.
Mbembe, Achille. "Provisional Notes on the Postcolony." *Africa* 62, no. 1 (1992): 3–37.
Payn, Howard. *The Merchandise Marks Act of 1887: With Special Reference to the Importation Sections and the Customs Regulations and Orders Made Thereunder together with the Conventions with Foreign States for Protection of Trade Marks and Orders in Council under the Patents, Designs, and Trade Marks Act, 1883, etc.* London, 1888.
Pritchard, Sara B. "From Hydroimperialism to Hydrocapitalism: 'French' Hydraulics in France, North Africa, and Beyond." *Social Studies in Science* 42, no. 4 (2012): 591–615.
Putuma, Koleka. *Collective Amnesia.* Cape Town: uHlanga, 2017.
Rediker, Marcus. "History from below the Water Line: Sharks and the Atlantic Slave Trade." *Atlantic Studies* 5, no. 2 (2008): 285–97.
Samuelson, Meg. "Coastal Form: Amphibian Positions, Wider Worlds, and Planetary Horizons on the African Indian Ocean Littoral." *Comparative Literature* 69, no. 1 (2017): 16–24.
Sandhu, Sukhdev. "Hydropoetics." environment.as.nyu.edu/docs/IO/25786/ENGL-UA973TopicsHydropoetics.pdf (accessed August 24, 2016).
Sheikh, Fariha. "The Alfred and the Open Sea: Periodical Culture and Nineteenth-Century Settler Emigration at Sea." *English Studies in Africa* 57, no. 1 (2014): 21–32.

Toward a Critical Ocean Studies for the Anthropocene

ELIZABETH DELOUGHREY

Abstract Recently, scholars have called for a "critical ocean studies" for the twenty-first century and have fathomed the oceanic depths in relationship to submarine immersions, multispecies others, feminist and Indigenous epistemologies, wet ontologies, and the acidification of an Anthropocene ocean. In this scholarly turn to the ocean, the concepts of fluidity, flow, routes, and mobility have been emphasized over other, less poetic terms such as blue water navies, mobile offshore bases, high-seas exclusion zones, sea lanes of communication (SLOCs), and maritime "choke points." Yet this strategic military grammar is equally vital for a twenty-first-century critical ocean studies for the Anthropocene. Perhaps because it does not lend itself to an easy poetics, the militarization of the seas is overlooked and underrepresented in both scholarship and literature emerging from what is increasingly called the blue or oceanic humanities. This essay turns to the relationship between global climate change and the US military, particularly the Navy, and examines Indigenous challenges to the militarism of the Pacific in the poetry of Craig Santos Perez.
Keywords blue humanities, Anthropocene, climate change, militarism, Pacific studies

While this special issue of *ELN* on "Hydro-criticism" was being written, the largest maritime exercise in history was taking place in the Pacific Ocean. Twenty-five thousand military personnel descended on the ocean area between the Hawaiian archipelago and Southern California to participate in "war games," including nearly fifty naval ships, two hundred aircraft, and five submarines. The twenty-sixth biennial Rim of the Pacific (RIMPAC) exercise comprised the military forces of twenty-five predominantly Pacific Rim nations, with the notable exceptions of China and Russia.[1] The theme of the five-week-long RIMPAC 2018 was "Capable, adaptive, partners"; its purpose, according to the US Navy, was to "demonstrate the inherent flexibility of maritime forces" in regard to everything from disaster relief to "sea control and complex warfighting."[2] Past war games had included exercises like sinking warships; this time the agenda listed amphibious operations, explosive ordnance disposal, mine clearance, and diving and salvage work, as well as the live firing of antiship and naval-strike missiles.[3] While US imperial interests in the region have categorized the largest ocean on our planet as an

57:1, April 2019 DOI 10.1215/00138282-7309655

Figure 1. US Nuclear Test *Swordfish*, Operation Dominic, 1962.

"American Lake," military incursion by the People's Republic of China into the Spratly and Paracel Islands has increased the Pentagon's concern that the Pacific is rapidly becoming a "Chinese Lake" and incentivizing military buildup in the region.[4]

Recently, scholars have called for a "critical ocean studies" for the twenty-first century and have fathomed the oceanic depths in relationship to submarine immersions, multispecies others, feminist and Indigenous epistemologies, wet ontologies, and the acidification of an Anthropocene ocean.[5] This is a welcome move after decades of scholarship that positioned the ocean as an anthropocentric and colonial "aqua nullius," or a blank space across which a diasporic masculinity might be forged.[6] In this new scholarship, an animated ocean has come into being as "wet matter" rather than inert backdrop.[7] In this recent scholarly turn to the ocean, the concepts of fluidity, flow, routes, and mobility have been emphasized over other, less poetic terms such as blue water navies, mobile offshore bases, high-seas exclusion zones, sea lanes of communication (SLOCs), and maritime "choke points." Yet this strategic military grammar is equally vital for a twenty-first-century critical ocean studies for the Anthropocene. Perhaps because it does not lend itself to an easy poetics, the militarization of the seas is overlooked and underrepresented in both scholarship and literature emerging from what is increasingly called the blue or oceanic humanities.

This is surprising, given that while the ocean may often be out of sight, the US Navy has long devoted its budgets to the visual reproduction of its military power at

Figure 2. The ships of the RIMPAC exercise, 2018.

sea, suggesting the mutual imbrication of technoscience and militarism. This includes the spectacular Cold War photography and films of nuclear weapons testing in the Marshall Islands (1946–62), which are widely available on YouTube. In figure 1 we see just one example of the visual display of a US naval vessel in direct relationship to the violent force of a nuclear weapon. Taken from Operation Dominic, where the US launched thirty-one nuclear weapons in the Pacific in the wake of the Bay of Pigs invasion, it shows the twenty-ton antisubmarine nuclear explosion named *Swordfish*, fired by the ship in the foreground, the *USS Agerholm*. In Paul Virilio's terms, this is the way in which "observation and destruction . . . develop at the same pace . . . so that every surface immediately became war's *recording* surface, its *film*."[8] As a mode of warfare, the US military's visual reproduction of its destructive power over sea and airspace—the global commons—continues today in its social media blitz about RIMPAC exercises, including the show of force in figure 2, ample online videos, and its Twitter feed (see fig. 2 and #ShipsofRIMPAC).

Although marine biologists may point out that "every breath we take is linked to the sea" and that planet Earth is in fact "a *marine* habitat,"[9] another kind of planetary metabolism is equally constitutive—American militarization of the oceans is foundational to maintaining the global energy supply that undergirds what some call the Capitalocene.[10] Over 60 percent of the world's oil supply is shipped by sea, and over 20 percent of the Pentagon's budget goes to securing it.[11] Securing the flow of oil has been a vital US naval strategy—not to say "mission"—since the 1970s.[12] Some have warned that there is a "dangerous feedback loop between war and global warming" because the Pentagon, in protecting its energy interests through extensive maritime and overseas base networks, estimated at over seven thousand, is the world's largest consumer of energy and the biggest institutional contributor to global carbon emissions.[13] This seems shocking because carbon emissions are regularly tied to citizen consumption rather than to military expansion.

The US Navy and its associated air force emit some of the dirtiest bunker and jet fuels to secure the passage of maritime oil transportation; this energy in turn is consumed and emitted by the military in rates disproportionate to any nation.[14] Not only is this fuel cycle common knowledge in military circles, but the Pentagon was exempted from all the major international climate accords *and* from domestic carbon emission legislation.[15] It should concern Anthropocene scholars and those in the emergent field of the energy humanities that "militarism is the most oil exhaustive activity on the planet."[16]

Transoceanic militarism—via sail, coal, steam, or nuclear-powered ships and submarines—has long been tied to global energy sources, masculinity, and state power. Hosted by the US Navy's Pacific Fleet since 1971, RIMPAC's oceanic war games have been a way to make visible what the nineteenth-century naval historian Alfred Thayer Mahan famously termed "the influence of sea power upon history." While Captain Mahan recognized the sea as a commons, and even as "the common birthright of all people," he spent his influential career advocating "the development of sea power," for the United States, which was critical to its nineteenth-century expansion to an "insular empire" from Puerto Rico to the Philippines.[17] Mahan's political influence helped convince US leadership of the importance of sea and wind currents in positioning Hawai'i as a vital naval base and coal-refueling station as well as a bulwark against China.[18] The 1898 annexations reflected the rise of American naval imperialism, where newly acquired colonies like Guam (Guåhan) were administered by the US Navy as if the island were a ship. A few years later islands and atolls like American Samoa were claimed as essential to fuel the US military and ruled by the Navy as coaling stations.[19] From the (illegal) US annexation of Micronesia in 1947, creating the Trust Territory of the Pacific Islands, to the current US practice of claiming permanent military exclusion zones on the high seas to test weapons—nowhere has this sea power been more apparent than in the world's largest ocean.[20]

The Pacific Ocean as defined by geographers covers one-third of the world's surface area (63 million square miles), but to the US military it extends all the way to the western coast of India, a nation that now participates in RIMPAC and represents the largest naval force in South Asia. Significantly, in the spring of 2018 the US military renamed its largest base, the Hawaiian-located Pacific Command, the "US Indo-Pacific Command" (USINDOPACOM) in recognition of its new maritime regime, which has expanded to 100 million square miles, or a stunning "fifty-two percent of the Earth's surface."[21] This unprecedented naval territorialism was almost entirely overlooked in the press and has not yet factored into any scholarly discussions of the Anthropocene or oceanic humanities.

In fact, this recent change in transoceanic hydro-politics has produced all kinds of material for cultural analysis, suggesting an interesting relationship between militarism and literary production (and consumption). The commander of the US Indo-Pacific Command, Admiral Phil Davidson, has recently posted a fascinating "professional development reading and movie list" on their website.[22] The book list includes titles one would expect from a military command, such as those about war histories and strategies, with a particular focus on cyberwar. Condoleezza Rice's (nonironically) titled book *Democracy* is on the reading list, which may not be surprising, but the appearance of the book *Athena Rising: How and Why Men Should Mentor Women*, certainly is. Female protagonists are central to a number of the novels, such as a women's coming-of-age story by the Japanese author Mitsuyo Kakuta and Michael Ondaatje's *Anil's Ghost*, which excavates the legacies of state-sponsored violence in Sri Lanka and Argentina. The movie list also includes some titles of interest to humanities scholars, particularly to postcolonialists: *Beats of No Nation*, a film about child soldiers in Africa based on Uzodinma Iweala's novel, and *Lion*, based on Saroo Brierly's memoir of Indo-Pacific adoption. There is certainly rich material to consider here in the making of transoceanic naval literacy, and the intersection of hydro-criticism with military hydro-politics.

Like the expansion into the Pacific Islands in the nineteenth century, the US Navy's inclusion of the Indian Ocean in its definition of the Pacific derives from strategies of energy security. There are five vital "sea lines of communication" (SLOCs) that connect both oceans through a lifeline of oil shipments from the Middle East: the Straits of Malacca, Hormuz, and Bab el-Mandeb, and the Suez and Panama Canals. According to the US Navy website, "RIMPAC is a unique training opportunity that helps participants foster and sustain the cooperative relationships that are critical to ensuring the safety of sea lanes and security on the world's oceans."[23] Because the majority of oil exports are over water, US energy policy has become increasingly militarized and secured by the Navy, the largest oceanic force on the planet. Scholars such as Michael Klare have characterized the US military since the 2003 Iraq war "as a global oil protection service, guarding pipelines, refineries, and loading facilities in the Middle East and elsewhere."[24] US Navy spokespeople readily admit that RIMPAC is an exercise in "power projection," a political and military strategy to use the instruments of state power quickly and effectively in widely dispersed locations far from the territorial state. Others might use the term transoceanic empire, with the recognition that much of this (often nuclear) power is also submarine. Fluidity, mobility, adaptability, and flux—all terms associated with neoliberal globalization regimes as well as the oceanic or blue humanities—are also key words and strategies of twenty-first-century maritime militarism.

Postcolonial scholars recognize that Cold War politics reshaped academic funding channels, training and hiring, the formulation of departments (such as area studies), and even their vocabularies. Thus when the US annexed territories in Micronesia and put them in the hands of the Navy, it made academic funding available to anthropologists, including Margaret Mead, to study Pacific Islander cultures.[25] The rise of a twenty-first-century oceanic humanities would benefit from an interrogation of how it may participate in, mitigate, or challenge larger strategic interests, examining how our current geopolitics shape academic discourse, not to say funding. Simon Winchester, writing in the early 1990s at the inception of globalization studies, described what he called "Pacific Rising," noting that this oceanic turn—following the logic of transnational capital—was "quite simply" about "*power*." And that power was represented, celebrated, and contested in the rise of globalization studies, Asia-Pacific studies, and Indigenous Pacific studies, fields largely informed by new models, epistemologies, and ontologies of the sea.[26]

While globalization studies of the late *twentieth century* emerged in relationship to the rise of transoceanic capital and its flows of "liquid modernity," to borrow from Zygmunt Bauman, we might raise the question as to how *twenty-first-century* articulations of an oceanic humanities and a turn to "hydro-criticism" might be informed by larger geopolitical and geontological (or sea-ontological) shifts.[27] Since the Obama era the United States has made a "Pacific pivot" that includes transoceanic militarism as well as a trade treaty that, according to Robert Reich, entails "forty percent of the world economy."[28] The Trans-Pacific Partnership (TPP)—critiqued as "NAFTA on steroids"— includes an attempt to solidify transnational energy and seabed mining interests over state environmental protections.[29] Of course, its key security agents are naval forces, particularly evident in the highly contested military "mega buildup" on Guåhan, one of the Navy's many "lily pads" and refueling stations, which some American pilots refer to as "the world's largest gas station."[30] In a remarkable erasure of Indigenous presence, many militarized islands and atolls of US-occupied Micronesia have been referred to as "unsinkable aircraft carriers" since the World War II era.[31] This is how militarized "ocean-space" is transformed into a "force-field," a term Philip E. Steinberg uses to describe the merging of the "ideological value of sea power" with "the key role of a strong 'blue-water' fleet in troop mobility, naval warfare" in the quest toward the "domination of distant lands."[32]

We might rightly turn the focus of *hydro-criticism* toward *hydro-power*, defined as energy, force, militarism, and empire. This raises the question as to the purpose of literary criticism in an era of expanding transoceanic militarism. Clearly it is no longer fashionable to publish literary anthologies celebrating the masculine heroic achievements of the Navy in verse, as it was for the British and Americans in the nineteenth and early twentieth centuries. But it should interest us that the largest military command on the planet is offering reading lists. As we turn to new sites of planetary expansion, flow, energy, and fluidity, we might ask, Where is the body of literature and scholarship responding to these global shifts in hydro-power? Where is the literary, artistic, and cultural critique of an aquatic territorialism of 52 percent of the Earth's surface?

Amitav Ghosh raises similar questions in tracing the relationship between energy, petrocapitalism, narrative, and the Anthropocene. In *The Great Derangement* he builds on his earlier observation that, given the ways in which the world economy is undergirded by oil, it is peculiar that there have been so few "petrofiction" novels. Twenty-five years later he asks why, in an era of disastrous climate change, we see so few literary responses that take on its global scope.[33] While he focuses exclusively (and problematically) on what he calls "literary fiction," I believe Ghosh's observations are relevant to calling attention to the lacuna in oceanic studies scholarship and literary production about US militarism more broadly.[34] Ghosh concludes that the European novel—which I would add was developed at the advent of an industrialism fueled by the labor and resources of the colonies—conceals "the exceptional" to promote "regularity" and thus naturalize bourgeois life.[35] This development narrowed the scale of what he terms "serious fiction" to an anthropocentric focus as well as a time scale that cannot account for the *longue durée*.[36] Thus, when faced with catastrophic climate change or nonhuman agency, the European-derived novel has difficulty engaging the "uncanny intimacy of our relationship with the nonhuman." He raises a provocatively maritime question: "Are the currents of global warming too wild to be navigated in the accustomed barques of narration?"[37]

Of course, no other region on the planet has been so deeply engaged with oceanic and maritime metaphors as Indigenous Pacific studies, which has drawn extensively on the image of the voyaging canoe as a vessel of the people and metaphor for navigating the challenges of globalization and ongoing colonialism.[38] Ghosh may had come to different conclusions if he had extended his analysis to Indigenous, feminist, and/or postcolonial fiction, which often challenge the human/nonhuman binary of western patriarchal thought and depict violence against non-European, nonnormative others as precisely that which prevents access to the "regularity" of bourgeois life. (In fact, his own novels might be considered as part of this postcolonial critique.) However, his analysis is particularly valuable for thinking about a history of silence and erasure when it comes to telling stories about the energies that undergird global capitalism—and, I would add, global militarism—in "the preserves of serious fiction."[39]

In the space I have remaining I want to turn to the Chamorro author Craig Santos Perez, who has written extensively about the voyaging-canoe metaphor in the wake of transoceanic militarism, and might be the only poet on the planet to turn to the RIMPAC exercises and inscribe their impact on both human and nonhuman ocean ecologies. While his medium is experimental poetry rather than the realist novel, his challenges to western binary thinking, the uniformity of traditional genre, and the separation of militarism from the transoceanic imaginary have much to say about decolonizing both genre and the broader Pacific, or Oceania.

Author of the multibook project *from unincorporated territory* (a reference to the political status of Guåhan), Perez is the winner of a PEN award and "imagines the blank page as an excerpted ocean, filled with vast currents, islands of voices, and profound depths."[40] Like other Indigenous poets from Oceania, a term Epeli Hau'ofa famously suggested as more representative of the flows of the region than the "Pacific," Perez has positioned his poetry as an oceanic vessel.[41] His work has plumbed the depths of an oceanic imaginary, particularly visible in his epic 2016 World Oceans Day "eco-poem-film" *Praise Song for Oceania*, in which he engages the ocean as origin, breath, body, mother, and absorber of plastic waste. In framing not just the permeability between humans and the ocean but their mutual responsibility and accountability, the speaker begs forgiveness for

> our territorial hands
> & acidic breath / please
> forgive our nuclear arms &
> naval bodies[42]

Drawing inspiration from a range of poets and scholars who've inscribed a transoceanic imaginary—including Hau'ofa—Perez concludes in praise of "our most powerful metaphor . . . / our trans-oceanic/past, present & future/flowing through our blood."[43] This embodied ocean, represented in the video through the sounds of breath and a heartbeat, foregrounds mergers between the human and a planetary nonhuman other that are naturalized (as breath, mother) and are also violent ("our nuclear arms").

Since the beginning of his *from unincorporated territory* series, (*[hacha]* in 2008), Perez has rendered visible a military that is too often "hidden in plain sight."[44] He has critiqued the history and depiction of Guåhan as a strategic naval base, as "USS *Guam*," and has framed his poems as "provid(ing) a strategic position for 'Guam' to emerge" from colonial and military hegemony. As such, he draws extensively on Indigenous voyaging traditions to poetically contest and mitigate the US Navy, reshaping what Ghosh has called the "accustomed barques of narration." The cover of *from unincorporated territory [saina]* (2010) juxtaposes a drawing of a Chamorro voyaging canoe, or sakman, above a photograph of the aircraft carrier USS *Abraham Lincoln* leading smaller naval ships in their patrol of the Indian Ocean in 2008 (fig. 3).[45] Although the world ocean has been partitioned into discrete national and international territories via the United Nations Law of the Sea (UNCLoS), the US Navy considers each of its aircraft carriers "four and a half acres of sovereign and mobile American territory."[46] Of course, the USS *Lincoln* is

Figure 3. The cover of *from unincorporated territory [saina]*, 2010. Courtesy of the author.

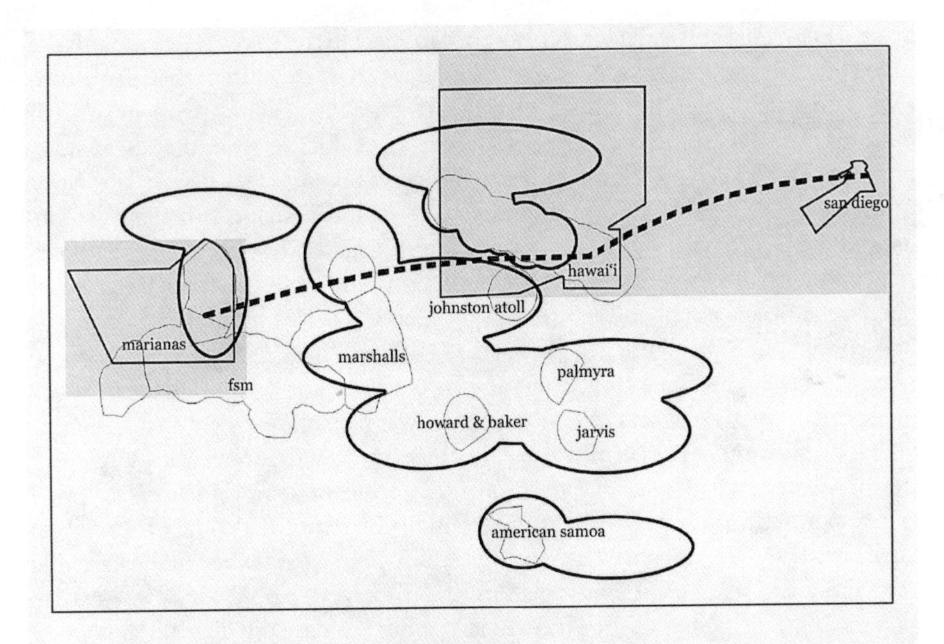

poemap based on the "Key US Bases in Pacific Pivot Buildup" map, prepared by Juan Wilson of Island Breath (www.islandbreath.org) (30 July 2014, Revision: 1.3.1). Sources: Map A: Mariana Island Training Area EIS (http://mitt-eis.com/), Map B: NOAA IOOS: Pacific Ocean States, territories, etc (http://www.ioos.noaa/gov/regions/pacioos.html), and Map C: DITC: Acoustic Effects on Marine Mammals (http://dtic.mil/dtic/tr/fulltext/u2/a560973.pdf) Page 3.

Figure 4. Poemap from *from unincorporated territory [lukao]*, 2017. Courtesy of the author.

the ship from which George W. Bush infamously declared, "Mission Accomplished," in May 2003 after the ship launched "16,500 sorties from its deck, and fired 1.6 million pounds of ordnance from its guns" the previous month during Operation Iraqi Freedom.[47] For complex reasons, Pacific Islanders continue to serve in disproportionate numbers in US military campaigns, lending nuance to the juxtaposition of these two maritime vessels of sovereignty in which Chamorro claims are tied to Indigenous sovereignty as well as US patriotism.[48]

Perez's four books of poetry engage US naval colonialism in Oceania, particularly in Guåhan, where the Navy occupies one-third of an island that is only thirty miles long.[49] In his most recent book, *from unincorporated territory [lukao]* (2017), he incorporates a number of "poemaps" that visualize military buildup and contamination in the region (fig. 4), as well as turning directly to the RIMPAC exercises of

2014. Thus Perez shifts the focus from the US Navy to the larger RIMPAC alliance of twenty-two nations, calling attention to the ways in which transnational militarism across the Indian and Pacific Oceans reflects a new era of hydro-politics. For example, the US military's "Cooperative Strategy for 21st Century Seapower" emphasizes a closer relationship between agencies, such as between the Navy, Coast Guard, and Marine Corps, as well as international alliances that are also evident in the US Department of Defense's "2014 Climate Change Adaptation Roadmap"; both call for a new era of HADR or Humanitarian Assistance/Disaster Relief operations because global warming is considered a threat multiplier.[50] It is well known that the commander of the largest US naval base (USPACOM), Admiral Locklear, in 2013 declared climate change the biggest security threat to the nation.[51] Since then US naval officers have argued for a "war plan orange for climate change," which involves more HADR operations in other countries because "these overtures may increase US access and these nations' receptiveness to hosting temporary basing or logistics hubs in support of future military operations."[52] Hence they call for larger RIMPAC activities, a 25 percent increase in ships sent to the Middle East, and by 2020 a 60 percent increase in ships and aircraft deployed to the new ocean known to the Pentagon as the "Indo-Asia-Pacific."[53] These are the military hydro-politics of the Anthropocene.

Perez's RIMPAC poem weaves together the fluid intimacy between mother and newborn daughter alongside the larger-scale militarism of Oceania. The poem is titled "(first ocean)," and its epigraph reads "*during the rim of the pacific military exercises, 2014*." It intersperses the Navy's ecological damage to all oceanic creatures—human and otherwise—with his newborn daughter's first immersion in the ocean. The use of parentheses in the poem's title invokes a placental or bodily enclosure of the infant, perhaps reminding the reader—like the conclusion of "Praise Song for Oceania"—that "our briny blood" connects us to the sea and our first placental ocean.[54]

The poem employs a second-person address (*you*) to his wife, highlighting familial intimacy. It traces out the baby's first introductions to water by her mother in Hawai'i, moving from being rinsed in the sink to taking a bath to becoming immersed in the sea. Each watery rinsing, bathing, and cleaning is juxtaposed to the repercussions of naval militarism: "pilot whales, deafened/by sonar" emerge "bloated and stranded/ashore." The speaker wonders "what will the aircrafts, ships, soldiers,/ and weapons of 22 nations take from [us]." In response we learn of the loss of the child's grandfather, whose ashes were "scattered in the pacific decades ago," as well as the death of "schools of recently spawned fish" that lie in the tidelands, "lifeless."[55] In this way the child meets both the body of her grandfather and the necropolitics of US militarism. These are multispecies mergers, but they are primarily about the military violence that undergirds Anthropocene extinctions. It has been widely reported that whale strandings and other animal deaths increase during and after RIMPAC exercises.[56] The poem concludes with a haunting question: "is Oceania memorial/or target, economic zone or monument/territory or mākua."[57] *Mākua*, the Hawaiian word for "parent," also refers to the highly contested military reservation at Mākua Valley on O'ahu, a place in Kanaka Maoli stories where humans originated, yet is where sacred Hawaiian sites and endangered species have been regularly bombed by the US military since the 1920s.[58] The poem calls attention to the ways in which the militarization of Oceania causes a rupture in the responsibilities of the *mākua* to the child, a rupture in the *kuleana*, or chain of responsibility, that connects all living beings and matter. The collection as a whole, by telescoping between the ordinary and the catastrophic, maternal intimacy and a militarized world ocean, brings together the very components that Ghosh notes are central to our understanding of the Anthropocene, yet so difficult to narrate in (western) prose. Perez demonstrates what Ghosh calls the "uncanny intimacy of our relationship with the nonhuman" and raises vital questions about intergenerational survival and responsibility.

The 2014 RIMPAC war games invoked by the poem led to the widespread devastation of marine wildlife and a 2015 ruling by a federal judge that the US Navy exercises, especially the use of explosives and sonar, were endangering millions of marine mammals.[59] The Navy's activities were harming over sixty populations of whales, dolphins, seals, and sea lions, and they "admitted that 2000 animals would be killed or permanently injured" by sonar or ship strike in the 2014 RIMPAC exercises.[60] This includes such species as endangered blue, fin, and beaked whales; false killer whales; spinner dolphins; melon-headed whales; and endangered Hawaiian monk seals. The court determined that there was a "breathtaking assertion" by the US Navy that their oceanic exercises "allow for no limitation at all," in terms of time, space, species, or depth, and that there was no justification for needing "continuous access to every single square mile of the Pacific."[61] Moreover, in a devastating—if not cleverly literary—ruling, Judge Susan Oki Mollway determined:

> Searching the administrative record's reams of pages for some explanation as to why the Navy's activities were authorized by the National Marine Fisheries Service ("NMFS"), this court feels like the sailor in Samuel Taylor Coleridge's "The Rime of the Ancient Mariner" who, trapped for days on a ship becalmed in the middle of the ocean, laments, "Water, water, every where, Nor any drop to drink."[62]

A critical ocean studies for the Anthropocene would bring together geopolitics with the literary and, like the poet Craig Santos Perez and federal judge Mollway, narrate them in ways that mutually inflect and inform each other. And hydro-criticism would be attentive to both hydro-politics and hydro-power. While the recent oceanic turn has produced scholarship that presses our understanding of the ontological and epistemological fluidity of our oceanic planet, a vigorous engagement with naval hydro-politics would help us better articulate and imagine a demilitarized future.

ELIZABETH DELOUGHREY is professor in English and in the Institute of the Environment and Sustainability at the University of California, Los Angeles. She is author of *Routes and Roots: Navigating Caribbean and Pacific Island Literatures* (2007) and coeditor of *Caribbean Literature and the Environment: Between Nature and Culture* (2005), *Postcolonial Ecologies: Literatures of the Environment* (2011), and *Global Ecologies and the Environmental Humanities: Postcolonial Approaches* (2015).

Acknowledgments

Mahalo nui to Anne Keala Kelly for first bringing my attention to the RIMPAC exercises and for her invaluable feedback on this article. Thanks also to my colleagues Keith Camacho and Victor Bascara for their ongoing demilitarization work and to Craig Santos Perez for sharing his insights and images for this article.

Notes

1 Brazil was invited but withdrew, reducing the number to twenty-five. China was "disinvited" due to its territorial expansion in the South China Sea. See *Maritime Executive*, "RIMPAC 2018 Begins."

2 *Navy News Service*, "U.S. Navy Announces." On this "new ecosecurity imaginary," see Robert P. Marzec's compelling argument about how neoliberal concepts of adaptation are "where ecosystems meet the war machine" (*Militarizing the Environment*, 2).

3 *Navy News Service*, "U.S. Navy Announces."

4 Hayes, Zarsky, and Bello, *American Lake*. On China, see Forsythe, "Possible Radar"; and Prabhakar, Ho, and Bateman, *Evolving Maritime Balance*.

5 "Critical ocean studies" and "sea ontologies" are explored in DeLoughrey, "Submarine Futures." This article is in conversation with important work by Steinberg and Peters, "Wet Ontologies"; Alaimo, *Exposed*; Helmreich, "Genders of Waves"; and Neimanis, *Bodies of Water*. See also Hessler, *Tidalectics*.

6 This is a larger argument taken up in relation to the British maritime (and shipwreck) fiction as well as more recent black Atlantic discourse in my *Routes and Roots*.

7 See Bélanger and Sigler, "Wet Matter."

8 Virilio, *War and Cinema*, 68.

9 NOAA biologist Nancy Foster quoted in Earle, *Sea Change*, xiv.

10 The term was first used by Andreas Malm and then further developed by Jason Moore and Donna Haraway. See Malm, *Fossil Capital*; Moore, *Capitalism in the Web of Life*; and Haraway, *Staying with the Trouble*.

11 Liska and Perrin, "Securing Foreign Oil."

12 In his 1974 address to the Rotary Club of San Francisco, Secretary of the Navy J. William Middendorf III argued: "It is the mission of the US Navy to protect the sea lanes for the transport of these critical [energy] imports. And it is the mission of the US Navy to render a political and diplomatic presence in the world today in support of our national policy" ("World Sea Power," 241).

13 Lawrence, "US Military Is a Major Contributor"; Hynes, "Military Assault on Global Climate"; Sanders, *Green Zone*. On the estimation of the number of US military bases (many of them top secret), see Johnson, *Nemesis*; and Lutz, *Bases of Empire*. While Lutz calculates at least 1,000 overseas bases, the US Department of Defense itself declares that it has "more than 7,000 bases, installations, and other facilities" ("2014 Climate Change Adaptation Roadmap," foreword).

14 Sanders, *Green Zone*.

15 The Pentagon was given an exemption from reporting its carbon emissions at the Kyoto Convention on Climate Change. See Hynes, "Military Assault on Global Climate"; and Neslen, "Pentagon to Lose Emissions Exemption."

16 Hynes, "Military Assault on Global Climate."

17 Mahan, *Influence of Sea Power*, 42, 43. On the history, see Thompson, *Imperial Archipelago*.

18 See Adomeit, "Alfred and Theodore Go to Hawai'i [*sic*]." Mahalo to Anne Keala Kelly for this reference and for her kokua regarding the naval history of Hawai'i.

19 US Navy rule of American Samoa from 1900 to 1951 catalyzed the Mau protests. See Chappell, "Forgotten *Mau*." My thanks to my colleague Keith Camacho for his insights on US Navy rule in the Pacific Islands. On resistance to militarism in the Pacific, see Shigematsu and Camacho, *Militarized Currents*. Walden Bello's article in that collection describes US presence in the Pacific as "a transnational garrison state that spans seven sovereign states and the vast expanse of Micronesia" (310) and points out that the US Navy was the main force behind the acquisition of Hawai'i, Guam/Guåhan, and the Philippines (315). On the Indigenous responses to American and Japanese militarism in the Marianas, see Camacho, *Cultures of Commemoration*; and Camacho, *Sacred Men*. For the Chamorro historical context, see Perez, *from unincorporated territory [hacha]*, preface.

20 On the weapons testing zones, see Van Dyke, "Military Exclusion and Warning Zones."

21 Wikipedia, "United States Indo-Pacific Command."

22 US Indo-Pacific Command, "Commander, US Indo-Pacific Command Professional Development Reading and Movie List."

23 "RIMPAC is the world's largest international maritime exercise" (US Navy, "RIMPAC 2014").

24 Klare, "Garrisoning the Global Gas Station."

25 Terrell, Hunt, and Gosden, "Dimensions of Social Life." This is discussed in DeLoughrey, *Routes and Roots*, 104–5.

26 Winchester, *Pacific Rising*, 27. Key texts that used the ocean as a trope for globalization include Connery, "Oceanic Feeling"; and Hau'ofa, "Our Sea of Islands." The edited collections that these essays appeared in—

Wilson and Dissanayake, *Global/Local*, and Wilson and Dirlik, *Asia/Pacific as Space of Cultural Production*—were critical to shifting US literary and cultural studies to the Pacific.

27 Bauman's liquid metaphors for globalization (in *Liquid Modernity*) are discussed in DeLoughrey, *Routes and Roots*, 225–26; and in Helmreich, *Sounding the Limits of Life*, 102. Elizabeth A. Povinelli has coined the term *geontologies* (in *Geontologies*), which I have developed into *sea ontologies* (in "Submarine Futures"). On wet ontologies, see Steinberg and Peters, "Wet Ontologies."

28 Reich, "Trans-Pacific Partnership."

29 Wallach, "NAFTA on Steroids"; Solomon and Beachy, "Dirty Deal."

30 Brooke, "Looking for Friendly Overseas Base." On "lily pads" and "forward operating locations," see Lutz, *Bases of Empire*, 20, 37. See also Natividad and Kirk, "Fortress Guam."

31 See Norris, "Air Assault on Japan," 86.

32 Steinberg, *Social Construction of the Ocean*, 17.

33 Ghosh, "Petrofiction." In *The Great Derangement* Ghosh argues that "if certain literary forms are unable to negotiate these torrents [of climate change chaos], then they will have failed—and their failures will have to be counted as an aspect of the broader imaginative and cultural failure that lies at the heart of the climate crisis" (8).

34 Ghosh, *Great Derangement*, 7.

35 Ghosh, *Great Derangement*, 17.

36 Ghosh, *Great Derangement*, 11, 59.

37 Ghosh, *Great Derangement*, 33, 8.

38 Hau'ofa, *We Are the Ocean*; Jolly, "Imagining Oceania"; Clifford, *Routes*. Diaz and Kauanui have argued that the "Pacific is on the move," in terms of tectonics, human migration, and a growing field of scholarship, in "Native Pacific Cultural Studies," 317; I have built on these works in *Routes and Roots*, which argues for a "transoceanic imaginary" (37). On the oceanic turn and its lack of engagement with Indigenous Pacific studies, see Somerville, "Where Oceans Come From."

39 Ghosh, *Great Derangement*, 11. I have in mind the Māori author Keri Hulme, whose poetry-fiction collection *Stonefish* imagines multiple scales for the Anthropocene. This is discussed in my "Submarine Futures" and expanded in *Allegories of the Anthropocene*.

40 *Lantern Review Blog*, "Page Transformed."

41 Hau'ofa, *We Are the Ocean*. The oceanic vessel metaphor in Pacific and black Atlantic literature is explored in greater depth in *Routes and Roots*.

42 Perez and Chong, *Praise Song for Oceania*. See also Perez, "Chanting Waters."

43 Hau'ofa argues that "the sea is our pathway to each other and to everyone else, the sea is our endless saga, the sea is our most powerful metaphor, the ocean is in us" (*We Are the Ocean*, 58).

44 Ferguson and Turnbull, *Oh, Say, Can You See*, xiii.

45 Perez, *from unincorporated territory [saina]*. The photo is drawn from Wikipedia Commons, commons.wikimedia.org/wiki/File:US_Navy_080905-N-7981E-845_The_aircraft_carrier_USS_Abraham_Lincoln_(CVN_72)_leads_a_formation_of_ships_from_the_Abraham_Lincoln_Strike_Group_they_transit_the_Indian_Ocean.jpg.

46 See Lutz, *Bases of Empire*, 4.

47 Sanders, *Green Zone*, 60.

48 See the essays collected in Bascara, Camacho, and DeLoughrey, "Gender and Sexual Politics of Pacific Island Militarisation."

49 See Camacho and Monnig, "Uncomfortable Fatigues," 158.

50 Mabus et al., "Cooperative Strategy for 21st Century Seapower"; US Department of Defense, "2014 Climate Change Adaptation Roadmap."

51 Bender, "Chief of US Pacific forces."

52 McGeehan, "War Orange for Climate Change." On how the US military is using climate change to incentivize expansion, see Marzec, *Militarizing the Environment*.

53 Mabus et al., "Cooperative Strategy for 21st Century Seapower."

54 Earle, *Sea Change*, 15.

55 Perez, "(first ocean)," in *from unincorporated territory [lukao]*, 17.

56 Fergusson, "Whales Beware."

57 Perez, "(first ocean)," 17.

58 Activist groups such as Mālama Mākua and EarthJustice have brought the military to court to halt the bombing, at least for the time being. On the militarism of Hawai'i and Mākua in particular, see Anne Keala Kelly's powerful film *Noho Hewa: The Wrongful Occupation of Hawai'i*. See also Carter, "'Let's Bomb This!'"

59 *EarthJustice*, "Court Rules Navy Training in Pacific Violates Laws."

60 Natural Resources Defense Council, "Navy Agrees to Limit Underwater Assaults on Whales and Dolphins."

61 Conservation v. National Marine Fisheries, 6876 (Hawai'i District Court, 2013). earthjustice.org/sites/default/files/files/2013-12-16NAVYSonarComplaint.pdf.

62 Conservation v. National Marine Fisheries, Civ. No. 13-00684 SOM/RLP, March 31, 2015; Natural Resources v. National Marine Fisheries, Civ. No. 14-00153 SOM/RLP (Hawai'i District Court, 2015). earthjustice.org/sites/default/files/files/2015-3-31%20Amended%20Order.pdf.

Works Cited

Adomeit, Ambjörn L. "Alfred and Theodore Go to Hawai'i [*sic*]: The Value of Hawai'i in the Maritime Strategic Thought of Alfred Thayer Mahan." *International Journal of Naval History* 13, no. 1 (2016). www.ijnhonline.org/2016/05/26/alfred-and-theodore-go-to-hawaii-the-value-of-hawaii-in-the-maritime-strategic-thought-of-alfred-thayer-mahan/#fn-1869-1.

Alaimo, Stacy. *Exposed: Environmental Politics and Pleasures in Posthuman Times.* Minneapolis: University of Minnesota Press, 2016.

Bascara, Victor, Keith L. Camacho, and Elizabeth DeLoughrey, eds. "Gender and Sexual Politics of Pacific Island Militarisation." Special issue. *Intersections: Gender and Sexuality in Asia and the Pacific*, no. 37 (2015). intersections.anu.edu.au/issue37_contents.htm.

Bauman, Zygmunt. *Liquid Modernity.* Cambridge: Polity, 2000.

Bélanger, Pierre, and Jennifer Sigler, eds. "Wet Matter." Special issue. *Harvard Design Magazine*, no. 39 (2014). harvarddesignmagazine.org/issues/39.

Bello, Walden. "Conclusion: From American Lake to a People's Pacific in the Twenty-First Century." In *Militarized Currents: Toward a Decolonized Future in Asia and the Pacific*, edited by Setsu Shigematsu and Keith L. Camacho, 309–23. Minneapolis: University of Minnesota Press, 2010.

Bender, Bryan. "Chief of US Pacific Forces Calls Climate Biggest Worry." *Boston Globe*, March 9, 2013. www.bostonglobe.com/news/nation/2013/03/09/admiral-samuel-locklear-commander-pacific-forces-warns-that-climate-change-top-threat/BHdPVCLrWEMxRe9IXJZcHL/story.html.

Brooke, James. "Looking for Friendly Overseas Base, Pentagon Finds It Already Has One." *New York Times*, April 7, 2004. www.nytimes.com/2004/04/07/us/looking-for-friendly-overseas-base-pentagon-finds-it-already-has-one.html.

Camacho, Keith L. *Cultures of Commemoration: The Politics of War, Memory, and History in the Mariana Islands.* Honolulu: University of Hawai'i Press, 2011.

Camacho, Keith L. *Sacred Men: Law, Torture, and Retribution in Guam.* Durham, NC: Duke University Press, forthcoming.

Camacho, Keith L., and Laurel A. Monnig. "Uncomfortable Fatigues: Chamorro Soldiers, Gendered Identities, and the Question of Decolonization in Guam." In *Militarized Currents: Toward a Decolonized Future in Asia and the Pacific*, edited by Setsu Shigematsu and Keith L. Camacho, 147–80. Minneapolis: University of Minnesota Press, 2010.

Carter, Keala. "'Let's Bomb This!': US Army Wants to Resume Live-Fire Training in Sacred Hawaiian Valley." *Intercontinental Cry*, March 6, 2018. intercontinentalcry.org/lets-bomb-this-us-army-wants-to-resume-live-fire-training-in-sacred-hawaiian-valley.

Chappell, David A. "The Forgotten *Mau*: Anti-Navy Protest in American Samoa, 1920–1935." *Pacific Historical Review* 69, no. 2 (2000): 217–60. doi.org/10.2307/3641439.

Clifford, James. *Routes: Travel and Translation in the Late Twentieth Century.* Cambridge, MA: Harvard University Press, 1997.

Connery, Christopher L. "The Oceanic Feeling and the Regional Imaginary." In *Global/Local: Cultural Production and the Transnational Imaginary*, edited by Rob Wilson and Wimal Dissanayake, 284–311. Durham, NC: Duke University Press, 1996.

DeLoughrey, Elizabeth. *Allegories of the Anthropocene.* Durham, NC: Duke University Press, 2019.

DeLoughrey, Elizabeth. *Routes and Roots: Navigating Caribbean and Pacific Island Literatures.* Honolulu: University of Hawai'i Press, 2007.

DeLoughrey, Elizabeth. "Submarine Futures of the Anthropocene." *Comparative Literature* 69, no. 1 (2017): 32–44. doi.org/10.1215/00104124-3794589.

Diaz, Vicente M., and J. Kēhaulani Kauanui. "Native Pacific Cultural Studies on the Edge." *Contemporary Pacific* 13, no. 2 (2001): 315–41. scholarspace.manoa.hawaii.edu/bitstream/handle/10125/13574/v13n2-315-342.pdf?sequence=1.

Earle, Sylvia A. *Sea Change: A Message of the Oceans.* New York: Putnam, 1995.

EarthJustice. "Court Rules Navy Training in Pacific Violates Laws Meant to Protect Whales, Sea Turtles." April 1, 2015. earthjustice.org/news/press/2015/court-rules-navy-training-in-pacific-violates-laws-meant-to-protect-whales-sea-turtles.

Ferguson, Kathy E., and Phyllis Turnbull. *Oh, Say, Can You See: The Semiotics of the Military in Hawai'i.* Minneapolis: University of Minnesota Press, 1998.

Fergusson, Manjari. "Whales Beware." *Hawaii Independent*, August 22, 2014. hawaiiindependent.net/story/whales-beware.

Forsythe, Michael. "Possible Radar Suggests Beijing Wants 'Effective Control' in South China Sea." *New York Times*, February 23, 2016. www.nytimes.com/2016/02/24/world/asia/china-south-china-sea-radar.html.

Ghosh, Amitav. *The Great Derangement: Climate Change and the Unthinkable.* Chicago: University of Chicago Press, 2016.

Ghosh, Amitav. "Petrofiction: The Oil Encounter and the Novel." *New Republic*, no. 206 (1992): 29–34.

Haraway, Donna J. *Staying with the Trouble: Making Kin in the Cthulucene*. Durham, NC: Duke University Press, 2016.

Hau'ofa, Epeli. "Our Sea of Islands." In *Asia/Pacific as Space of Cultural Production*, edited by Rob Wilson and Arif Dirlik, 86–100. Durham, NC: Duke University Press, 1995.

Hau'ofa, Epeli. *We Are the Ocean: Selected Works*. Honolulu: University of Hawai'i Press, 2008.

Hayes, Peter, Lyuba Zarsky, and Walden Bello. *American Lake: Nuclear Peril in the Pacific*. New York: Viking, 1986.

Helmreich, Stefan. "The Genders of Waves." *WSQ* 45, nos. 1–2 (2017): 29–51. doi.org/10.1353/wsq.2017.0015.

Helmreich, Stefan. *Sounding the Limits of Life: Essays in the Anthropology of Biology and Beyond*. Princeton, NJ: Princeton University Press, 2015.

Hessler, Stefanie, ed. *Tidalectics: Imagining an Oceanic Worldview through Art and Science*. Cambridge, MA: MIT Press, 2018.

Hulme, Keri. *Stonefish*. Wellington: Huia, 2004.

Hynes, H. Patricia. "The Military Assault on Global Climate." *Truthout*, September 8, 2011. truthout.org/articles/the-military-assault-on-global-climate.

Johnson, Chalmers. *Nemesis: The Last Days of the American Republic*. New York: Holt, 2008.

Jolly, Margaret. "Imagining Oceania: Indigenous and Foreign Representations of a Sea of Islands." In *Framing the Pacific in the Twenty-First Century: Co-existence and Friction*, edited by Daizaburo Yui and Yasua Endo, 29–48. Tokyo: Center for Pacific and American Studies, University of Tokyo, 2001.

Klare, Michael T. "Garrisoning the Global Gas Station." *Global Policy Forum*, June 12, 2008. www.globalpolicy.org/component/content/article/154-general/25938.html.

Lantern Review Blog. "The Page Transformed: A Conversation with Craig Santos Perez." March 12, 2010. www.lanternreview.com/blog/2010/03/12/the-page-transformed-a-conversation-with-craig-santos-perez.

Lawrence, John. "The US Military Is a Major Contributor to Global Warming." *San Diego Free Press*, November 14, 2014. sandiegofreepress.org/2014/11/the-us-military-is-a-major-contributor-to-global-warming.

Liska, Adam J., and Richard K. Perrin. "Securing Foreign Oil: A Case for Including Military Operations in the Climate Change Impact of Fuels." *Environment: Science and Policy for Sustainable Development*, July–August 2010. www.tandfonline.com/doi/abs/10.1080/00139157.2010.493121.

Lutz, Catherine. *The Bases of Empire: The Global Struggle against U.S. Military Posts*. New York: New York University Press, 2009.

Mabus, Ray, et al. "A Cooperative Strategy for 21st Century Seapower." US Marine Corps, Navy, and Coast Guard, March 2015. www.navy.mil/local/maritime/150227-CS21R-Final.pdf.

Mahan, A. T. *The Influence of Sea Power upon History, 1660–1783*. New York: Dover, 1987.

Malm, Andreas. *Fossil Capital: The Rise of Steam Power and the Roots of Global Warming*. London: Verso, 2016.

Maritime Executive. "RIMPAC 2018 Begins, but without China." June 29, 2018. www.maritime-executive.com/article/rimpac-2018-begins-but-without-china#gs.hahJMZ8.

Marzec, Robert P. *Militarizing the Environment: Climate Change and the Security State*. Minneapolis: University of Minnesota Press, 2016.

McGeehan, Timothy. "A War Plan Orange for Climate Change." *U.S. Naval Institute Proceedings Magazine*, October 2017. www.usni.org/magazines/proceedings/2017-10/war-plan-orange-climate-change.

Middendorf, J. William. "World Sea Power: U.S. vs. U.S.S.R." In *Oceans: Our Continuing Frontier*, edited by H. William Menard and Jane L. Schieber, 238–43. Del Mar, CA: Publisher's, 1976.

Moore, Jason W. *Capitalism in the Web of Life: Ecology and the Accumulation of Capital*. London: Verso, 2015.

Natividad, LisaLinda, and Gwyn Kirk. "Fortress Guam: Resistance to US Military Mega-Buildup." *Asia-Pacific Journal* 8, no. 19 (2010): 1–17. apjjf.org/-LisaLinda-Natividad/3356/article.html.

Natural Resources Defense Council. "Navy Agrees to Limit Underwater Assaults on Whales and Dolphins." Press release, September 14, 2015. www.nrdc.org/media/2015/150914.

Navy News Service. "U.S. Navy Announces Twenty-Sixth Rim of the Pacific Exercise." May 30, 2018. www.navy.mil/submit/display.asp?story_id=105789.

Neimanis, Astrida. *Bodies of Water: Posthuman Feminist Phenomenology*. London: Bloomsbury, 2017.

Neslen, Arthur. "Pentagon to Lose Emissions Exemption under Paris Climate Deal." *Guardian*, December 14, 2015. www.theguardian.com/environment/2015/dec/14/pentagon-to-lose-emissions-exemption-under-paris-climate-deal.

Norris, John G. "The Air Assault on Japan." *Flying*, November 1943, 21–23, 86.

Perez, Craig Santos. "Chanting the Water." 2016. craigsantosperez.com/chanting-waters-2016.
Perez, Craig Santos. *from unincorporated territory [hacha]*. Kaneohe, HI: Tinfish, 2008.
Perez, Craig Santos. *from unincorporated territory [lukao]*. Oakland, CA: Omnidawn, 2017.
Perez, Craig Santos. *from unincorporated territory [saina]*. Oakland, CA: Omnidawn, 2010.
Perez, Craig Santos, and Justyn Ah Chong. *Praise Song for Oceania*. YouTube video, June 14, 2017. craigsantosperez.com/praise-song -oceania.
Povinelli, Elizabeth A. *Geontologies: A Requiem to Late Liberalism*. Durham, NC: Duke University Press, 2016.
Prabhakar, Lawrence W., Joshua H. Ho, and Sam Bateman. *The Evolving Maritime Balance of Power in the Asia-Pacific: Maritime Doctrines and Nuclear Weapons at Sea*. Singapore: Institute of Defence and Strategic Studies, 2006.
Reich, Robert. "Robert Reich on Trans-Pacific Partnership as 'NAFTA on Steroids.'" YouTube video, posted by Business and Human Rights Resource Centre, January 29, 2015. www .business-humanrights.org/en/robert-reich-on -trans-pacific-partnership-as-nafta-on-steroids.
Sanders, Barry. *The Green Zone: The Environmental Costs of Militarism*. Chico, CA: AK, 2009.
Shigematsu, Setsu, and Keith L. Camacho, eds. *Militarized Currents: Toward a Decolonized Future in Asia and the Pacific*. Minneapolis: University of Minnesota Press, 2010.
Solomon, Ilana, and Ben Beachy. "A Dirty Deal: How the Trans-Pacific Partnership Threatens Our Climate." Washington, DC: Sierra Club, 2015. content.sierraclub.org/creative-archive /sites/content.sierraclub.org.creative-archive /files/pdfs/1197%20Dirty%20Deals% 20Report%20Web_03_low.pdf.
Somerville, Alice Te Punga. "Where Oceans Come From." *Comparative Literature* 69, no. 1 (2017): 25–31. doi.org/10.1215/00104124-3794579.
Steinberg, Philip. *The Social Construction of the Ocean*. Cambridge: Cambridge University Press, 2001.
Steinberg, Philip, and Kimberley Peters. "Wet Ontologies, Fluid Spaces: Giving Depth to Volume through Oceanic Thinking." *Environment and Planning D: Society and Space* 33, no. 2 (2015): 247–64. doi.org/10.1068/ d14148p.
Terrell, John Edward, Terry L. Hunt, and Chris Gosden. "The Dimensions of Social Life in the Pacific: Human Diversity and the Myth of the Primitive Isolate." *Current Anthropology* 38, no. 2 (1997): 155–95. doi.org/10.1086/204604.
Thompson, Lanny. *Imperial Archipelago: Representation and Rule in the Insular Territories under U.S. Dominion after 1898*. Honolulu: University of Hawai'i Press, 2010.
US Department of Defense. "2014 Climate Change Adaptation Roadmap." Alexandria, VA: US Department of Defense, 2014. www.acq.osd .mil/eie/downloads/CCARprint_wForward_e .pdf.
US Indo-Pacific Command. "Commander, US Indo-Pacific Command Professional Development Reading and Movie List." n.d. www.pacom.mil /About-USPACOM/COM-PACOM -Professional-Development-Reading-List (accessed December 3, 2018).
US Navy. "RIMPAC 2014." www.cpf.navy.mil /rimpac/2014 (accessed August 3, 2018).
Van Dyke, Jon M. "Military Exclusion and Warning Zones on the High Seas." In *Freedom for the Seas in the Twenty-First Century: Ocean Governance and Environmental Harmony*, edited by Jon M. Van Dyke, Durwood Zaelke, and Grant Hewison, 445–70. Washington, DC: Island, 1993.
Virilio, Paul. *War and Cinema: The Logistics of Perception*. London: Verso, 1989.
Wallach, Lori. "NAFTA on Steroids." *Nation*, June 27, 2012. www.thenation.com/ article/nafta-steroids.
Wikipedia. "United States Indo-Pacific Command." en.wikipedia.org/w/index.php?title=United_ States_Indo=Pacific_Command&oldid =851229307 (accessed August 3, 2018).
Winchester, Simon. *Pacific Rising: The Emergence of a New World Culture*. New York: Prentice-Hall, 1991.

The Oceanic South

MEG SAMUELSON AND CHARNE LAVERY

Abstract This essay proposes the category of the oceanic South. It presents the Southern Hemisphere's blue expanses as one of its defining features and elaborates from this a framework that brings into agitated contention the extractive economies of the North, the persistent legacies of settler colonialism in the South, and other interlocking human and more-than-human itineraries. Tracking a drift into the Southern Ocean in the fiction of J. M. Coetzee, the essay takes this "most neglected of oceans" as a vantage point from which to draw the contours of the oceanic South and engage its troubled surfaces and lively depths. Thinking through the roiling and hostile, fecund, and unbounded nature of this ocean, the essay follows "the lives of whales" in novels by Witi Ihimaera and Zakes Mda. Sounding the ocean's imaginative depths, these fictions offer illuminating ways of thinking the South while maintaining an unsettling planetarity.
Keywords critical ocean studies, the South, *Elizabeth Costello*, *The Whale Caller*, *The Whale Rider*

The southern region of the globe is most readily conceived of as what is bound by the longitudinal lines of imperial and metropolitan domination or described by the curvier Brandt line as comprising the "poorer nations."[1] But it might also be defined by the relatively vast maritime expanses that distinguish the Southern Hemisphere: 80 percent of its surface area is composed of seas and oceans, compared to 60 percent of the North. In emphasizing the oceanic composition of a hemisphere that is "south in more ways than one,"[2] we seek to situate these conceptual grids in a fluid and lively framework that is both "three-dimensional and turbulent."[3] This framework brings into agitated contention the extractive economies of the North and the persistent legacies of settler colonialism in the South, along with the questions of intrahuman and more-than-human justice that flow between them. It registers the wakes of south-south connectivity that have been drawn across bodies of water to grant substance to normative imaginings of the global South[4] while impelling attention to the ocean as a "material space of nature."[5]

We bring the oceanic South into focus by tilting the conventional axis of southern thought to center Antarctica and its encircling Southern Ocean.[6] This is to perform a gesture of what Gayatri Spivak calls "planetarity," which is to inhabit

57:1, April 2019 DOI 10.1215/00138282-7309666

the planet as "the species of alterity" by effecting the "defamiliarization of familiar space."[7] While the sea is essentially alien to humans,[8] the Southern Ocean is particularly so. It is the only body of water in which waves circulate without encountering intervening landmasses, thus growing gargantuan in size and ferocity. Even though the velocity of its winds connected the three land-bound oceans by the express Clipper Route during the age of sail and of empire, the roaring, furious, and screaming latitudes remain daunting to maritime traffic. The Antarctic Convergence—where icy currents meet warmer sub-Antarctic waters and of which there is no northern equivalent—supports an abundance of marine life, but, like the icebound continent itself, it has not nested human habitation.[9]

The nature of this ocean—simultaneously roiling and hostile, fecund and unbounded—suggests ways of redrawing the contours of the South and engaging both its troubled surfaces and its lively depths. Because it uniquely flows into the Atlantic, Indian, and Pacific Oceans, the Southern Ocean opens up possibilities for tracking the intersecting currents and itineraries that compose the oceanic South. It enables us to sketch out the oceanic South as a category that draws together the dispersed landmasses of the settler South, the decolonized and still colonized South, the "sea of islands" comprising Indigenous Oceania,[10] and the frozen continent of Antarctica. Within this category, the geographic and material conditions of the Southern Hemisphere are interfused with the "normative force" of a postcolonial or global South that dreams of new worlds to come and likewise with the numinous worlds of the Indigenous South.[11] At the same time, this tumultuous roadstead fogs up any aspiration to conceptual clarity. Our initial attempt to elaborate the oceanic South from the defamiliarizing perspective that it offers is thus necessarily, but also in principle, tentative and even wavering. Establishing the Southern Ocean as a vantage point for articulating "southern theory" or "theory from the South," we forgo both the stability of being "grounded" and the lucidity of a privileged position on global processes in favor of indicating ways of reading that are alive to turbulence, drift, refraction, and the more-than-human materialities of the oceanic South.[12]

"The Most Neglected of Oceans"

Fiction by the South African–Australian author J. M. Coetzee leads us into the Southern Ocean in appropriately digressive and disorienting ways. Both the novels that Coetzee wrote under the states of emergency in late apartheid South Africa—*Foe* (1986) and *Age of Iron* (1990)—imagine drifting "southward" to the "bleak winter-waters of the most neglected of oceans."[13] Cast as inhuman and ahistorical, the Southern Ocean appears to offer a retreat from the tyranny of the times. But in these works, as well as in the more extended reflection on this ocean that is presented in the later *Elizabeth Costello: Eight Lessons* (2003), it is instead shown to induce unsettling reckonings.

Foe reprises the maritime mercantile triangle that Daniel Defoe's *Robinson Crusoe* (1719) had traced through the points of England, South America, and West Africa. Shifting from the horizontal to the vertical axis, its conclusion stages a dive into the depths in which the St. Lucian poet Derek Walcott has located a "subtle and submarine" alternative to the history contained in the "monuments and martyrs" of the colonizing world.[14] Many readers have followed Coetzee's narrator into these warm waters to sift through the murky sediment accumulating on the seafloor and investigate the contents of the wrecked ship. We wish instead to track the rowing boat from which the narrator slips overboard. Abandoned on the surface, the boat "bobs away, drawn south toward the realm of whales and eternal ice."[15] This transient scene is elaborated in Coetzee's subsequent novel, *Age of Iron*. Sickened by the inferno she has witnessed in the militarized townships of apartheid, and soiled by her complicity in that state, the narrator, Mrs. Curren, expresses a longing to cast off from that land and sail to the "latitudes where albatrosses fly," and there to be "lash[ed] . . . to a barrel or a plank" and left "bobbing on the waves under the great white wings."[16]

The movement in both of these scenes is indicated in the reiterated action of "bobbing." Simultaneously an up-and-down motion and an uncharted lateral drift, it is an action that floats vertiginously away from the conceptual anchor provided by the ship. In *Foe* this is the vessel of the triangular trade that commodifies human life and erases biodiversity to prop up the global North with profits extracted from southern plantations that were cultivated with slave labor; in *Age of Iron*, it is the "worm-riddled" "sinking ship" of the late apartheid state.[17] Both narratives certainly address themselves to these mercantile and geopolitical structures, but the undulating movement of the two scenes also unmoors them from such conceptual frameworks. Drifting into the realm of the whales and beneath the circumpolar itineraries of the wandering (or white-winged) albatross, they offer an orientation on the planet that—by virtue of remaining all at sea—is both defamiliarizing and confounding.[18] The bobbing action that is so pronounced in the heaving Southern Ocean reminds us that the sea is not simply "a space that *facilitates* movement—the space across which things move—but . . . is a space that is *constituted by* and *constitutive of* movement."[19] This inherent motility encourages a drifting and unsettled mode of reading, whether of the text or of the world.

In her inquiry into the concept of "world," Kelly Oliver observes that the etymology of *planet* "is from the Greek *planetes*, meaning 'wanderer.'"[20] Asking what it might mean to consider ourselves "wanderers on this wandering planet that is our home," Oliver refers to Julia Kristeva's expression that "we are strangers to ourselves" and locates the possibilities of the "struggle for social justice" in this roving and unhomely state.[21] The condition of this exposure to planetarity is implicit in what Paul Gilroy has recently described as a "lowly watery orientation" that resists the "high altitude theorizing" of universalizing conceptions of the Anthropocene that smooth over racialized fractures in the category of the human.[22] Gilroy turns to the Mediterranean to elaborate his proposed alternative figure of an "offshore humanism," but he approaches this figure by first evoking Herman Melville's *Moby-Dick*, with its "passionate planetary ontology," and the "unsettling story" of *Benito Cereno*.[23] Both works point us toward the oceanic South as offering a generative fetch for "sea level theory."

Mrs. Curren's drift into the Southern Ocean in *Age of Iron* also embodies "sea level theory." Rather than presenting northern traffic in southern waters, it conveys the unsettling implications of the Southern Hemisphere. In a telling exchange, Coetzee responds to a reading of *Age of Iron* as offering "absolution" to its author-narrator Mrs. Curren (whom Coetzee here calls Elizabeth) by describing the novel as "more troubled (in the sense that the sea can be troubled)."[24] The reference is presumably to Isaiah 57:20: "But the wicked are like the troubled sea, when it cannot rest, whose waters cast up mire and dirt." Mrs. Curren's imagined recourse to the Southern Ocean does thus not acquit her of the barbarities in which she is complicit by virtue of her situation in the apogee of settler colonialism that was apartheid South Africa. Instead, the albatross that wanders into this novel from Coleridge's *Rime* suggests the interminable atonement that the violent betrayal of southern hospitality demands.[25] Reflecting on what the proper relation to the history of that betrayal should be on the part of "a representative of the generation in Africa for whom apartheid was created," Coetzee has tendered the "dubious and hesitant" response of "liv[ing] out the question" in his writing.[26]

Coetzee elaborates this question in the Southern Ocean in "The Novel in Africa," the second "lesson" (or chapter) of *Elizabeth Costello*, which ranks among the most neglected works in Coetzee's oeuvre. The eponymous protagonist of *Elizabeth Costello* is—like Mrs. Curren—an authorial figure. The scenario of the second lesson is that Costello, the "famous Australian writer," is appointed to deliver a lecture on the "The Future of the Novel" as part of the entertainment and education program of a cruise ship, the SS *Northern Lights*, as it makes passage across the Southern Ocean from Christchurch to Cape Town via the Ross Ice Shelf of Antarctica. Sandwiched between her address and a subsequent talk billed as "'The Lives of Whales,' with sound recordings" is the lecture on "The Novel in Africa," which is presented by the Nigerian novelist Emmanuel Egudu.[27] Following the drift of Coetzee's earlier allusion to Coleridge, *Elizabeth Costello* churns interspecies justice into the question that is "lived out" in its pages, while the exchange between an African and an Australian writer on the *Northern Lights* stages an encounter between the global South and the hemispheric South that remains unresolved in this troubled narrative.

Costello declares herself drawn to the voyage by the prospect of visiting Antarctica and "feel[ing] what it is like to be a living, breathing creature in spaces of inhuman cold."[28] Egudu, in contrast, having been displaced by the underdevelopment of the global South and the northern slant of the global literary marketplace, is on a perpetual lecture circuit. The dichotomy that is apparently set up on the Southern Ocean is between the shared creaturely life of this planet and the uneven distribution of global goods that determines Egudu's position. This dichotomy is, however, neither shored up nor dissolved in the narrative; instead, its binary terms are left fluctuant. The eight lessons of *Elizabeth Costello*, after all, chart Costello's own peripatetic journeys on the global lecture circuit: in the first lesson, during which she delivers a lecture on realism, Costello has cause to respond acerbically to the North American presumption that she inhabits the "far edges" of the globe while describing herself as one of the "late settlers" on that "vast" continent who are "only fleas on Australia's backside";[29] and the final lesson brings her to account for colonial "atrocities" in Tasmania,[30] the southernmost state of the "South Land."

Whereas Costello oscillates between the Southern Hemisphere and the global North, Egudu voices the more determined critique that issues from the global South in his lecture "The Novel in Africa." Addressing the *Northern Lights*' wealthy passengers, who style themselves as "ecological tourists," Egudu draws attention to the "global system" in which "it has been allotted to Africa to be the home of poverty."[31] Egudu remarks that his audience (who hail from both hemispheres but enjoy membership in the endowed North) are cruising toward "one of the remoter corners of the globe," while quoting Paul Zumthor on how "Europe has spread across the world . . . ravag[ing] life forms, animals, plants, habitats, languages." Indicting the world-destroying and planet-deranging appetites of the North, his lecture extols the enlivening breath of African orality. This, too, is proffered for consumption on the cruise ship. The entertainment program of the *Northern Lights* presents a microcosm of the global market in which Egudu suffers exoticization while being in turn required to peddle his alterity to remain in circulation. "Even here," Egudu concludes, "on this ship sailing towards the continent that ought to be the most exotic of all, and the most savage, the continent with no human standards at all, I can sense I am exotic."[32] Costello, who comments caustically on the seductions of exoticism, nonetheless admits that Egudu's is "the one black face in this sea of white."

The dazzling whiteness of the icy ocean in which his "black face" appears "exotic" reflects the invasive and devouring histories of European expansion into the Southern Hemisphere and the construction of Antarctica as the "last frontier" of imperialism.[33] As she dwells on this ocean, Costello recalls Edgar Allan Poe's description of strange Antarctic natives with dark skin and black teeth. Elizabeth Leane accounts for Poe's portrayal by noting that, "for English-speaking novelists, the proximity of Maori peoples to the Antarctic was clearly suggestive and often became one ingredient in an incoherent jumble of exoticized images of indigeneity that were drawn upon in populating the southern continent."[34] Responding perhaps to this allusion in his short story "Meeting Elizabeth Costello," the Maori writer Witi Ihimaera inserts his alter ego, called Wicked, into the scene on the *Northern Lights*. Adding to the entertainment program, Wicked delivers a lecture on "The Indigenous Novel in Antarctica" in lieu of his planned presentation on the "Maori Eden."[35] The lecture that he does not present would have celebrated "the lyricism and life-affirming literature of his own people," supported by clips from the film *Whale Rider*—a reference to the film adaptation of Ihimaera's earlier novel, which was greeted with a rapturous reception in the North and to which we turn in the next section.[36] Instead, Wicked delivers what Costello describes as a "rant."[37] But, after establishing a counternarrative that would seem to anchor alterity, Ihimaera's story concludes with another confounding turn as the familiar figure of Costello is refracted across a series of mirrors as if scattered across the icy surface of the sea.

Whereas the ocean through which the *Northern Lights* plows reflects an unrelieved whiteness, its depths are shown to be both vivid and more-than-human. During the dinner that follows Costello's and Egudu's lectures in "The Novel in Africa," the conversation turns from the problem of exoticism to the "tiny beings, tons of them to the square mile, whose life consists in being swept in serene fashion through these icy waters, eating and being eaten, multiplying and dying, ignored by history."[38] Waking early as the ship approaches Macquarie Island, Costello peers over the rail to observe a sea that is "alive" with penguins, "large, glossy-backed fish that bob and tumble and leap in the swell."[39] Here the bobbing action that we have highlighted in *Foe* and *Age of Iron* plunges beneath the surface to open up a three-dimensional, multispecies view. From this perspective, the agitation of the ocean is enlivening. These lively waters have, however, not been "ignored by history." The polite dining-room scene convened on the *Northern Lights* is premised on the histories of consumption that have buoyed up the global North and that fuel this pleasure cruise. When Costello recognizes that the seething mass of life she contemplates from the deck is composed of a raft of penguins, she recalls that Macquarie Island was a nineteenth-century hub of the penguin-oil industry and that thousands of penguins were clubbed to death and boiled down on these shores. This "most neglected of oceans" is thus rendered into the "story of global resource colonialism" that has constituted and incorporated the South in its world-historical form.[40]

"'The Lives of Whales,' with Sound Effects"

Mentioned but not represented in "The Novel in Africa" is the slated lecture "The Lives of Whales." The next two chapters of *Elizabeth Costello* instead feature the protagonist's celebrated—and controversial—addresses on "the lives of animals."[41] Rather than accompanying Costello to those lectures, delivered in the North, we maintain our focus on the oceanic South and take up the invitation to sound its depths with whales. "The Novel in Africa" issues one plumb line for this sounding in its reference to the penguin-oil industry, the precursor to whale-oil extraction in the Southern Ocean. Another line—one offered by the unrepresented lecture "The Lives of Whales" and its promised "sound effects"—is attuned to the loud and lively ocean. Both situate the oceanic South at the conjuncture of global and planetary forces and of human and more-than-human histories. We follow them by turning to two novels produced from the countries between which the *Northern Lights* cruises: Witi Ihimaera's *The Whale Rider* from Aotearoa New Zealand and Zakes Mda's *The Whale Caller* from South Africa.

Much has been written about the ways in which these novels stage achieved or failed interspecies relations. Rather than further rehearsing these relations, our interest lies in locating them in a set of intersecting itineraries that map out the oceanic South. What we wish to underline is how the itineraries of humans and whales "*interlock*" as they journey together "through the southern seas."[42] This is the term that Ihimaera uses to describe "the knowledge of whalespeaking" that the ancients once had, and with which the Maori ancestor Paikea asked a whale to carry him to the land that lay "far to the south."[43] The novel shows how the interlocking of "land inhabitants and ocean inhabitants" that articulates the origin story of Aotearoa is sundered when "the whalekilling beg[ins]."[44]

The whale oil extracted from southern waters "lubricated the wheels of industry" and illuminated the lamps of the post-Enlightenment North that are alluded to in the name of Coetzee's cruise ship, and that we receive as referring to both the civilizing discourses of the imperial North and the wedge of reason with which it cleaved apart the concepts of "nature" and "culture."[45] In his reading of *Moby-Dick*, the environmentalist literary scholar Lawrence Buell notes that whaling evolved into "an extractive industry of global scope" during the nineteenth century.[46] As the Artic hunting grounds were exhausted, this industry shifted increasingly southward. In the early decades of the twentieth century, approximately 2 million whales were killed in the Southern Hemisphere in "'a slaughter' . . . 'that has few parallels in the history of wildlife exploitation'" and which led to "the near-extinction of the great baleen whales."[47] This fin de siècle turn is anticipated in the "Southern whaling voyage" on which Melville's *Pequod* embarks: before it is upended by the white whale, it is bound to round "both stormy Capes" and the orientation of its trajectory is signaled by Ishmael's declaration, "Cetus is a constellation in the South!"[48] The violence of the "whalekilling" projected also onto the shores of the Indigenous South; as Ishmael observes in *Moby-Dick*, Australia "was given to the enlightened world by the whaleman."[49]

Formally manifesting the interlocking itineraries rent by the "whalekilling" that ensued off southern shores, Ihimaera's *The Whale Rider* alternates between first-person narration by the character Rawini and omniscient narration focalized by an ancient whale bull and presented in italics. Each traces out their respective routes at and below sea level. Rawini travels from Aotearoa New Zealand through urban Australia to a settler plantation in Papua New Guinea. The geography he traverses references histories of Pacific "blackbirding," land appropriation, habitat destruction, and persistent settler racisms. During this odyssey, says Rawini, "I grew into an understanding of myself as a Maori."[50] Essential to this understanding is his recognition of his kinship with Papua New Guineans, in what Alice Te Punga Somerville has identified as the novel's significant, and often ignored, articulation of Maori "connection with the Pacific."[51] This articulation locates the story in what Rawini describes as the "huge seamless marine continent which we call Te Moana Nui a Kiwi, the Great Ocean of Kiwa" that extends from Hawai'i in the north to Rapa Nui (or Easter Island) in the east and Aotearoa New Zealand in the south.[52]

The second narrative moves across Te Moana Nui a Kiwi (or what is variously called Oceania, the South Seas, and the South Pacific) and the Southern Ocean as it follows a whale herd in its migrations between "the cetacean crib" in the Valdes Peninsula of Patagonia, the Tuamotu Archipelago of the South Pacific, and Antarctica. The whales' journey along this ancient circuit is abruptly obstructed by nuclear testing in the Tuamotu Archipelago.[53] Human history collapses—or rather *erupts*—into natural history in the form of a "*scald[ingly]*" "*bright light*," and "*giant tidal soundwaves*" tear through the ocean and the creatures that inhabit it.[54] The whales' focalization sounds the depths of the ensuing devastation, enabling the novel to register the "*hairline factures indicating serious damage below the crust of the earth*."[55] Apprehending the danger of "*undersea radiation*," the cetacean protagonists break their migratory pattern "*to seek before time the silent waters of the Antarctic*."[56] The ferocity of the Southern Ocean is precisely what renders it a haven to the whales: a "*calm and unworldly*" realm is maintained beneath the "*inhuman, raging storm*" that sweeps across its surface.[57] But the urgency of the whales' flight has driven them farther south than ever before, and they come up against a solid wall of ice. During their disoriented retreat, they strand themselves on the shores of Whangara.

The ancient bull returns to this shore out of nostalgia for the "*oneness*" he had experienced with Paikea, the ancestral whale rider who had journeyed with him from the place of the gods to settle these islands a thousand years before. Koro, the aged descendant of Paikea who seeks to translate this legacy into the future, is at the time returning from presenting his claim in a land dispute hearing. Tellingly, the airplane on which he travels is described as "bucking like an albatross" before the storm that has blown in from Antarctica.[58] Human and whale itineraries then converge on the shore as Koro, his son Rawini, and his granddaughter Kahu attempt to save the beached herd. Both narratives conclude redemptively as Kahu fulfills Paikea's legacy and reactivates the state of "*interlock*" to steer the herd back to the ocean depths before disengaging to return to land. But the shore remains haunted by the deaths of the "two hundred members of a vanishing species" that beached the previous day at nearby Wainui, foreshadowing the arrival of the mythical herd and underlining the planet-deranging effects of the "*giant tidal soundwaves*" that the whales' sounding has exposed.[59]

Mda's novel *The Whale Caller* sketches out similarly intersecting itineraries of land and ocean inhabitants from the vantage point of the southern tip of Africa. The Whale Caller had learned the songs of migrating whales during his own peregrinations in which he "spent many years walking westwards along the coast of the Indian Ocean, until he reached the point where the two oceans met, and then proceeded northwards along the Atlantic Ocean coast."[60] His journey traces out a triangular structure that positions the Cape as the tempestuous portal of the northbound trade that plied between the Atlantic and Indian Oceans.[61] To the whales making their annual passage "from the sub-Antarctic to the warmth of South African waters,"[62] in contrast, the Cape offers a safe harbor in the Southern Ocean to which they repair after feeding in the krill-rich waters of the Antarctic Convergence. Their offshore presence inclines the story toward southern and submarine positions.

The narrative centers on the fishing village of Hermanus, which has developed into a popular whale-watching site due to its geophysical situation: the seafloor drops away steeply to provide cetaceans with a deep and sheltered bay for breeding, while the coastal cliffs furnish wildlife tourists with a natural amphitheater. Rather than enabling another manifestation of what *The Whale Rider* describes as "*interlock*," these tourists are depicted as the latest crest in a set of invasive waves to have washed up on the cliffs that the novel declares to have been once the "home" of the Indigenous Khoikhoi and San, who suffered the first wave of genocidal violence during the colonial invasion.

The Whale Caller surmises that—like the Australasians he has read about—the Indigenous inhabitants of these shores had feasted on stranded whales and that their expressions of gratitude for the bounty delivered by the sea included also mourning for the loss of companion species.[63] Once the site of ambivalent celebratory scenes convened around whale carcasses, the southern littoral subsequently became an unequivocal theater of death. The Whale Caller can still discern the "two-hundred-year-old stench from the slaughter of the southern rights by French, American and British whalers." Even though now a protected species, his beloved southern rights continue to hazard various anthropogenic dangers on their migratory route, including "pirates and poachers," "ships' propellers," "fishing gear entanglements and explosives from oil exploration activity."[64] When the sea retaliates against this violation by hurling its "black . . . rage" against the town of Hermanus, the whale Sharisha is unable to "distinguish the blue depths from the green shallows" to which she is lured by the Whale Caller blowing on his homemade kelp horn. His obsession with her has become consuming and he recognizes too late that, while he can summon Sharisha, he lacks the song to save her: unlike in the Aboriginal Dreamtime stories of which he has read, he cannot "sing to make the female whale and her calf escape the shallow waters," thus ensuring the survival of "future generations that would replenish the seas."

Following his disastrous failure to appreciate the condition of "*interlock*" that—as *The Whale Rider* registers in its conclusion—allows for communion across species without dissolving the difference between creatures of the land and those of the sea, the Whale Caller throws away his horn and commits to living out his days as a penitent. He thus returns us to the figure of the Ancient Mariner—as does "The Novel in Africa," which concludes with Costello observed by an indifferent albatross and her wary chick as if to suggest both the inhuman extent of the Southern Ocean and the imperiling of "future generations" that the human presence therein portends. After Coleridge's Ancient Mariner shoots an albatross in the Southern Ocean, he finds himself drifting "Alone, alone, all, all alone, / Alone on a wide wide sea" in the South Pacific.[65] He thus establishes the oceanic South as the setting for the emergent condition that Edward O. Wilson calls—rather than the Anthropocene—"The Age of Loneliness." The Mariner had earlier claimed that his ship was "the first that ever burst / Into that silent sea."[66] Yet, when Ihimaera's ancient whale rider and bull traverse southern seas to found Aotearoa, the ocean throbs with welcoming songs. Ihimaera thus redirects the Ancient Mariner's claim from historical priority in the oceanic South to responsibility for its silence. The violent intrusions of the North, rather than the *Anthropos*, is held accountable for despoiling the blue planet.[67]

Retelling an origin story of the Indigenous South and reactivating its legacy for the present, *The Whale Rider* presents ocean crossings that are world opening rather than world destroying, and soundings that stand in contrast to those of submarine nuclear explosions. This is demonstrated when the new whale rider and the ancient bull dive together into a roisterous world of "dolphin chatter, krill hiss, squid thresh, shark swirl, shrimp click and, ever present, the strong swelling chords of the sea's constant rise and fall."[68] At the same time, it maintains the inhumanness of the ocean: the whale rider must return to shore. Mda's Whale Caller similarly cannot meet Sharisha on shared ground, even as the welcome that Coleridge's albatross offers the mariners when they enter the frozen waters of the South is echoed in what he understands as their mutual "yearning" for one another—just as it is in the "yearning" that Ihimaera's earth and oceans feel for a human presence. The ethical injunction issued by the materiality of these seas, which—in Costello's thoughts—are "full of things that seem like us but are not," is thus to respect alterity in the scene of oceanic hospitality.[69]

Conclusion

Jean Comaroff and John Comaroff open their inquiry into "theory from the South" by quoting Costello's question to Egudu in "The Novel in Africa": "How can you explore a world in all its depth if at the same time you are having to explain it to outsiders?"[70] The problem that they identify in the global knowledge economy is precisely this relegation of southern positions to the "performance of otherness."[71] *Theory from the South* intervenes in this economy by arguing for the centrality and prescience of the South in ways that we have found to be highly generative. Reading from the oceanic South, however, offers different perspectives on unfathomed depths and the condition of alterity. As Hester Blum points out, "A critical stance emerging from the perspective of the sea should be mindful of registering the volumes of what its geophysical properties render inaccessible."[72]

In approaching the South from the Southern Ocean, we have sought to encounter its oceanic nature in heightened form: the three-dimensional motility and turbulence that are characteristic of seas in general are notably amplified in this ocean, which is defined by its unbounded reaches and relentless circulations. It too has been subject to the invasion and plunder that have produced the global South, but it also enables us, in Spivak's terms, to "imagine ourselves as planetary creatures rather than global entities," and thus to conclude that "alterity remains underived from us; it is not our dialectical negation, it contains us as much as it flings us away."[73]

The fiction considered in this article moves us toward resonant apprehensions as it sounds the oceanic South. It registers the persistent presence and legacies of northern resource extraction and settler colonialism in the South while being simultaneously alive to the more-than-human materialities of the sea and its inhabitants. As such, it illuminates ways of thinking between globe and planet, and of interlocking questions of intrahuman and more-than-human justice that are of growing import in a time of anthropogenic climate change.[74] The oceanic South suggests, finally, also a model of textuality and modes of reading appropriate to it. Its more-than-human materialities manifest in the agitated surfaces of texts that maintain the condition of alterity in an unsettling planetarity that is realized because of—rather than despite—their admittance of depths that elude and even repulse comprehension.

MEG SAMUELSON is associate professor in the Department of English and Creative Writing at the University of Adelaide and associate professor extraordinary at Stellenbosch University. Her recent research engages with coastal form in narrative fiction from the African Indian Ocean littoral, photography in Zanzibar, surfing cultures and the Indian Ocean shore break, sharks as uncanny figures of racial terror in the Anthropocene, ways of telling the China-in-Africa story in the Anthropocene, the southern orientations of J. M. Coetzee's writing, and the literary world ocean. She coedits the Palgrave series *Maritime Literature and Culture*.

CHARNE LAVERY is a researcher based at the Wits Institute for Social and Economic Research, University of the Witwatersrand, on a project called "Oceanic Humanities for the Global South." Her book *Remapping the World: Indian Ocean Fiction in English* is under revision, and she coedits the Palgrave series *Maritime Literature and Culture.*

Acknowledgments

Early versions of these ideas were presented at the "Sustaining the Seas" conference at the University of Sydney in December 2017, and as a joint paper at the WISH seminar series at the Wits Institute of Social and Economic Research (WiSER), Wits, Johannesburg in May 2018. We are grateful to participants at the conference and the seminar, to the anonymous reviewers for helpful feedback on an earlier version of this article, and to Sharad Chari for his particularly suggestive commentary.

Notes

1 The Brandt line measures GDP per capita and meanders along the thirty-degree parallel north before looping up to include Mongolia and China and then swerving around Japan, Singapore, Australia, and New Zealand; see Prashad, *Poorer Nations.*

2 We borrow this formulation from Isabel Hofmeyr's recent description of the South Atlantic in "Southern by Degrees," 81.

3 Steinberg and Peters, "Wet Ontologies, Fluid Spaces," 247.

4 For an exemplary study of transoceanic connectivity as a way of composing the global South, see Hofmeyr, "The Black Atlantic Meets the Indian Ocean."

5 Steinberg, *Social Construction*, 209.

6 The collaborative formulation of the "oceanic South" draws on our respective ongoing projects of "thinking from the Southern Ocean" (Lavery) and of elaborating "the blue southern hemisphere" and the writing of "southern worlds" in works by J. M. Coetzee and others (Samuelson).

7 Spivak, *Death of a Discipline*, 72, 77.

8 See Helmreich, *Alien Ocean.*

9 For an account of the Southern Ocean that is attentive to the material conditions of its history, see Antonello, "Southern Ocean."

10 Hau'ofa, *New Oceania.*

11 See Prakash, *Darker Nations*; and Cheah, *What Is a World?*

12 On southern theory as a theory that is "grounded" in the South, see Connell, *Southern Theory*; on "theory from the South" as a privileged vantage point on global processes, see Comaroff and Comaroff, *Theory from the South.*

13 Coetzee, *Age of Iron*, 108.

14 Walcott, "The Sea Is History."

15 Coetzee, *Foe*, 155.

16 Coetzee, *Age of Iron*, 108.

17 Coetzee, *Age of Iron*, 117.

18 See also Lavery, "Drift."

19 Steinberg, "Of Other Seas," 165.

20 Oliver, *Earth and World*, 103.

21 Oliver, *Earth and World*, 103.

22 Gilroy, "Offshore Humanism," 8; for a study that traces this fault line in marine settings, see Samuelson, "Thinking with Sharks."

23 Gilroy, "Offshore Humanism," 3.

24 Coetzee, *Doubling the Point*, 250.

25 See J. R. Ebbatson's proposal that the Mariner's act of shooting an albatross with a crossbow "may be a symbolic rehearsal of the crux of colonial expansion, the enslavement of native peoples; and that the punishments visited upon the Mariner, and the deaths of his shipmates because of their complicity, may represent European racial guilt, and the need to make restitution" ("Coleridge's Mariner," 198). Ebbatson supports his reading by detailing the young Coleridge's involvement in the abolitionist movement and his readings of travel narratives, including those of James Cook's southern voyages and, possibly, that of Vasco da Gama, which relates how his predecessor, Bartholomew Dias, shot and killed a native of what is assumed to be Mossel Bay in present-day South Africa with a crossbow (Ebbatson, "Coleridge's Mariner," 177, 179; Ravenstein, *Journal*, 10).

26 Attwell, "An Exclusive Interview with J. M. Coetzee."

27 Coetzee, *Elizabeth Costello*, 37–38.

28 Coetzee, *Elizabeth Costello*, 35.

29 Coetzee, *Elizabeth Costello*, 15, 29.

30 Coetzee, *Elizabeth Costello*, 202.

31 Coetzee, *Elizabeth Costello*, 49, 41.

32 Coetzee, *Elizabeth Costello*, 47.

33 Leane, *Antarctica in Fiction*, 50; the "Antarctic Club" of nations was exclusively "white" even though membership was extended to the southern states of Argentina, Chile, Australia, New Zealand, and—albeit to a significantly

lesser extent—apartheid South Africa (Hofmeyr, "Southern by Degrees," 83).

34 Leane, *Antarctica in Fiction*, 25.

35 Ihimaera, "Meeting Elizabeth Costello," 94, 93.

36 Ihimaera, "Meeting Elizabeth Costello," 93.

37 Ihimaera, "Meeting Elizabeth Costello," 100.

38 Melville, *Moby-Dick*, 248; Coetzee, *Elizabeth Costello*, 49.

39 Coetzee, *Elizabeth Costello*, 54.

40 Hofmeyr, "Southern by Degrees," 83.

41 The two chapters comprising "The Lives of Animals"—"The Philosophers and the Animals" and "The Poets and the Animals"—were first presented in Coetzee's Tanner lectures at Princeton University in 1997; "The Novel in Africa" was first delivered as the Una lecture at the University of California, Berkeley in 1998.

42 Ihimaera, *Whale Rider*, chap. 14.

43 Ihimaera, *Whale Rider*, chap. 7.

44 Ihimaera, *Whale Rider*, chap. 7.

45 Roberts, *Unnatural History*, 87, 95.

46 Buell, *Writing for an Endangered World*, 205. For a study of the historical continuum between whale-oil culture and petroculture, see Scott, "Whale Oil Culture."

47 Biologists Scott Baker and Phil Clapham, quoted in Roman, *Whale*, 142–43; Antonello, "Southern Ocean," 302.

48 Melville, *Moby-Dick*, 129, 99.

49 Melville, *Moby-Dick*, 97; Kim Scott's revisionist account of the so-called friendly frontier of southwestern Australia in *That Deadman Dance* centers whaling as both the activity through which settler colonialism extended into the region and a symbol of the laying waste of Indigenous and nonhuman worlds alike.

50 Ihimaera, *Whale Rider*, chap. 11.

51 Somerville, *Once Were Pacific*, 63.

52 Ihimaera, *The Whale Rider*, chap. 6.

53 Between 1966 and 1996 France performed repeated nuclear tests in the Mururoa Atoll of the Tuamota Archipelago, which remains one of its overseas territories.

54 See Dipesh Chakrabarty's signal observation that the Anthropocene names a caesura in which "the distinction between human and natural histories—much of which had been preserved even in environmental histories that saw the two entities in interaction—has begun to collapse" ("Climate," 207).

55 Ihimaera, *Whale Rider*, chap. 9.

56 Ihimaera, *Whale Rider*, chap. 9.

57 Ihimaera, *Whale Rider*, chap. 14.

58 Ihimaera, *Whale Rider*, chap. 16.

59 Ihimaera, *Whale Rider*, chap. 9; see DeLoughrey, *Roots and Routes*, 227: the novel "offers a broad and profoundly historical genealogy of Pacific peoples while also illuminating the ways in which all creatures that constitute the Pacific whakapapa are impacted by nuclear pollution."

60 Mda, *Whale Caller*, 9.

61 See Samuelson, "Rendering the Cape-as-Port"; Melville's Ishmael notably compares the Cape to "some noted four corners of a great highway, where you meet more travelers than in any other part" (Melville, *Moby-Dick*, 218).

62 Mda, *Whale Caller*, 16.

63 See Haraway, "Companion Species Manifesto."

64 Mda, *Whale Caller*, 13, 37.

65 Coleridge, *Rime*, pt. 4.

66 Coleridge, *Rime*, pt. 2.

67 See Ebbatson's reading of *The Rime*, which surmises that the "gigantic historical process" enacted in the shooting of the albatross includes western imperial expansion into native lands, the transoceanic slave trade, and the "massive slaughter of seals and whales" that followed Cook's reported wildlife sightings in the Antarctic regions ("Coleridge's Mariner," 180).

68 Ihimaera, *Whale Rider*, chap. 19.

69 Coetzee, *Elizabeth Costello*, 54.

70 Coetzee, quoted in Comaroff and Comaroff, *Theory from the South*, 2–3.

71 Comaroff and Comaroff, *Theory from the South*, 3.

72 Blum, "Melville and Oceanic Studies," 30.

73 Spivak, *Death of a Discipline*, 72–73.

74 See Chakrabarty, "Climate," 213: "The task of placing, historically, the crisis of climate change thus requires us to bring together intellectual formations that are somewhat in tension with each other: the planetary and the global; deep and recorded histories; species thinking and critiques of capital."

Works Cited

Antonello, Alessandro. "The Southern Ocean." In *Oceanic Histories*, edited by David Armitage, Alison Bashford, and Sujit Sivasundaram, 296–318. Cambridge: Cambridge University Press, 2018.

Attwell, David. "An Exclusive Interview with J. M. Coetzee." *Dagens Nyheter*, December 8, 2003. www.dn.se/kultur-noje/an-exclusive-interview-with-j-m-coetzee.

Blum, Hester. "Melville and Oceanic Studies." In *The New Cambridge Companion to Herman Melville*, edited by Robert S. Levine, 22–36. New York: Cambridge University Press, 2014.

Buell, Lawrence. *Writing for an Endangered World: Literature, Culture, and Environment in the U.S. and Beyond*. Cambridge, MA: Harvard University Press, 2001.

Chakrabarty, Dipesh. "The Climate of History: Four Theses." *Critical Inquiry* 35, no. 2 (2009): 197–222.
Cheah, Pheng. *What Is a World? On Postcolonial Literature as World Literature.* Durham, NC: Duke University Press, 2016.
Coetzee, J. M. *Age of Iron.* London: Secker and Warburg, 1990.
Coetzee, J. M. *Doubling the Point: Essays and Interviews,* edited by David Attwell. Cambridge, MA: Harvard University Press, 1992.
Coetzee, J. M. *Elizabeth Costello: Eight Lessons.* London: Secker and Warburg, 2003.
Coetzee, J. M. *Foe.* London: Penguin, 1986.
Coleridge, Samuel Taylor. *The Rime of the Ancient Mariner.* 1798. www.poets.org/poetsorg/poem/rime-ancient-mariner (accessed February 20, 2018).
Comaroff, Jean, and John L. Comaroff. *Theory from the South; or, How Euro-America Is Evolving toward Africa.* New York: Routledge, 2015.
Connell, Raewyn. *Southern Theory: The Global Dynamics of Knowledge in Social Science.* Sydney: Allen and Unwin, 2007.
Defoe, Daniel. *The Life and Strange Surprising Adventures of Robinson Crusoe of York, Mariner.* London, 1719. eBooks@Adelaide, 2014.
DeLoughrey, Elizabeth M. *Routes and Roots: Navigating Caribbean and Pacific Island Literatures.* Honolulu: University of Hawai'i Press, 2007.
Ebbatson, J. R. "Coleridge's Mariner and the Rights of Man." *Studies in Romanticism* 11, no. 3 (1972): 171–206.
Gilroy, Paul. "'Where Every Breeze Speaks of Courage and Liberty': Offshore Humanism and Marine Xenology; or, Racism and the Problem of Critique at Sea Level." *Antipode* 50, no. 1 (2018): 3–22.
Haraway, Donna. *The Companion Species Manifesto: Dogs, People, and Significant Otherness.* Chicago: Prickly Paradigm, 2003.
Hau'ofa, Epeli. *A New Oceania: Rediscovering Our Sea of Islands.* Suva: School of Social and Economic Development, University of the South Pacific in association with Beake House, 1993.
Helmreich, Stefan. *Alien Ocean: Anthropological Voyages in Microbial Seas.* Berkeley: University of California Press, 2009.
Hofmeyr, Isabel. "The Black Atlantic Meets the Indian Ocean: Forging New Paradigms of Transnationalism for the Global South—Literary and Cultural Perspectives." *Social Dynamics* 33, no. 2 (2007): 3–32.
Hofmeyr, Isabel. "Southern by Degrees: Islands and Empires in the South Atlantic, the Indian Ocean, and the Sub-Antarctic World." In *The Global South Atlantic,* edited by Kerry Bystrom and Joseph R. Slaughter, 81–96. New York: Fordham University Press, 2018.
Ihimaera, Witi. "Meeting Elizabeth Costello." In *Get on the Waka: Best Recent Maori Fiction,* edited by Witi Ihimaera, 90–101. Auckland: Reed, 2007.
Ihimaera, Witi. *The Whale Rider.* North Shore: Raupo, 2008. Kindle.
Lavery, Charne. "Drift." In "Toxicity, Waste, and Detritus in the Global South: South Africa and Beyond," edited by Pamila Gupta and Gabrielle Hecht. Special issue, *Somatosphere,* December 14, 2017. somatosphere.net/2017/12/drift.html.
Leane, Elizabeth. *Antarctica in Fiction: Imaginative Narratives of the Far South.* Cambridge: Cambridge University Press, 2012.
Mda, Zakes. *The Whale Caller.* Johannesburg: Penguin, 2006.
Melville, Herman. *Moby-Dick; or, The Whale,* edited by Tony Tanner. Oxford: Oxford University Press, 1988.
Oliver, Kelly. *Earth and World: Philosophy after the Apollo Missions.* New York: Columbia University Press, 2015.
Prashad, Vijay. *The Darker Nations: A People's History of the Third World.* New York: New, 2007.
Prashad, Vijay. *The Poorer Nations: A Possible History of the Global South.* London: Verso, 2013.
Ravenstein, E. G. *A Journal of the First Voyage of Vasco da Gama, 1497–1499.* Cambridge: Cambridge University Press, 1898.
Roberts, Callum. *The Unnatural History of the Sea.* Washington, DC: Island, 2007.
Roman, Joe. *Whale.* London: Reaktion, 2006.
Samuelson, Meg. "Rendering the Cape-as-Port: Sea-Mountain, Cape of Storms/Good Hope, and Adamastor in Local-World Literary Formations." *Journal of Southern African Studies* 42, no. 3 (2016): 523–37.
Samuelson, Meg. "Thinking with Sharks: Racial Terror, Species Extinction, and Other Anthropocene Fault Lines." *Australian Humanities Review* 62 (2018). australianhumanitiesreview.org/2018/12/02/thinking-with-sharks-racial-terror-species-extinction-and-the-other-anthropocene-fault-lines.
Scott, Heidi. "Whale Oil Culture, Consumerism, and Modern Conservation." In *Oil Culture,* edited by Ross Barrett and Daniel Worden, 3–18. Minneapolis: University of Minnesota Press, 2014.
Scott, Kim. *That Deadman Dance.* London: Picador, 2010.

Somerville, Alice Te Punga. *Once Were Pacific: Māori Connections to Oceania*. Minneapolis: University of Minnesota Press, 2012.

Spivak, Gayatri Chakravorty. *Death of a Discipline*. New York: Columbia University Press, 2003.

Steinberg, Philip. "Of Other Seas: Metaphors and Materialities in Maritime Regions." *Atlantic Studies* 10, no. 2 (2013): 156–69.

Steinberg, Philip. *The Social Construction of the Ocean*. Cambridge: Cambridge University Press, 2001.

Steinberg, Philip, and Kimberley Peters. "Wet Ontologies, Fluid Spaces: Giving Depth to Volume through Oceanic Thinking." *Environment and Planning D: Society and Space* 33, no. 2 (2015): 247–64.

Walcott, Derek. "The Sea Is History." In *Collected Poems, 1948–1984*. New York: Farrar, 1986.

The Underwater Imagination

From Environment to Film Set, 1954–1956

MARGARET COHEN

Abstract The 1940s and 1950s were a revolutionary era in both access to and imagination of the underwater environment in the most modernized nations of the globe. This article examines new possibilities in the exhibition of the underwater environment in film thanks to new dive and filming technologies. It focuses on how the depths are portrayed as a setting in Richard Fleischer's *20,000 Leagues under the Sea* (1954), by the Disney studio, and Jacques-Yves Cousteau and Louis Malle's *The Silent World* (1956). In both films, conventions for portraying terrestrial settings are adapted to the resistant conditions of the aquatic environment. Such adaptation unleashes creativity in filmmaking and results in portrayals in which audiences are at once at home in this alien area of our planet and able to assimilate and enjoy its unprecedented vistas and savage life.
Keywords history of technology, film, underwater environment, blue humanities

The 1940s and 1950s were a revolutionary era in the imagination of the underwater environment for publics in the most modernized nations of the globe. The impetus stimulating this revolution derived from new access to the depths during the World War II era, developed by both Allied and Axis militaries, in conjunction with new discoveries in science and new innovations in technology. This new access would take off in the immediate post–World War II years and transform oceanography, marine biology, warfare, and communications. It also would revolutionize the exhibition of the undersea realm on land. New experiences of the undersea emerged across a range of technologies and media, from print culture to new forms of aquariums such as oceanariums, giving a more interactive experience of marine life. Photography and film gained new tools to show in vivid color hitherto unknown, diverse environments, from coral reefs to the deep sea. Indeed, in the new cultural configurations enabling the public to view the undersea, photography and film played an outsized role, given the authority of these media as documentations of reality in the twentieth century.[1] The contributions extended beyond transmitting new views of the undersea to shaping actively the public perception of a remote and forbidding planetary environment that few could experience, even with the takeoff of diving as a leisure practice during the 1950s. In the words of

ENGLISH LANGUAGE NOTES
57:1, April 2019 DOI 10.1215/00138282-7309677

Helen Rozwadowski, leading historian of the modern undersea from the nineteenth century into the present, "Imagination may, in fact, play a larger role in our perception of the ocean, especially its third dimension, than modern science."[2]

The history of how the undersea is represented in film during this era thus is a subject with political and social ramifications, as well as part of a more specialized area of media-studies analysis. Despite its interest, however, underwater film has yet to become a robust subject of critical inquiry. Nicole Starosielski, one of the first critics to draw together underwater films as a coherent set of artifacts, notes the marginalization of "underwater films . . . in . . . ecocinema studies." While, as she states, "a number of works . . . have begun to address the cultural representations of marine mammals," exemplified by Greg Mitman's discussion of dolphins in *Reel Nature* (2009), "these lines of inquiry have yet to develop into a sustained conversation on the cinematic construction of the underwater environment."[3] Franziska Torma, another pioneer in underwater film scholarship, agrees: "Interdisciplinary research on media, mobility and exploration has mainly covered land-based travel from the nineteenth to the twentieth centuries."[4] Since Starosielski, Torma, and Rozwadowski's work on the subject, the conversation has developed, including Jonathan Crylen's "The Cinematic Aquarium" (2015) and writings by Ann Elias on the representation of coral, notably her forthcoming *Coral Empire*.[5]

This article contributes to the conversation by looking at how filmmakers of the 1950s constructed the undersea as a cinematic setting once they had the technologies to film diverse environments and capture their beauties and terror in color. Filmmakers' artistry was shaped by technological affordances—what Crylen has called "enabling technologies"—that initiated the revolution in underwater film.[6] As Torma observes, "The most important innovation of all was the invention of autonomous diving technologies, rooted in military technologies." In her words, "Mobility devices give birth to the myth of the 'merman.'"[7] Modern scuba was vastly less dangerous and expensive than the earlier military CCOUBA used by J. E. Williamson, who was the first to film underwater and contributed the underwater sequences to Stuart Paton's silent *20,000 Leagues under the Sea* (1916).[8] Scuba had a dramatic impact on the ability of people to film and act in a toxic atmosphere. Other revolutionary technologies for film of this era included ways to compensate for the extreme darkness of the underwater environment lit only by light from the surface filtered through the dense atmosphere of water. Refraction together with the magnification inherent in underwater optics were additional challenges. Engineers developed waterhousing, lenses, and flash adjusting variously for the distortions of underwater, low-light conditions, and the impact of increased depth pressure on cameras. Warmer and more flexible insulation than helmet diving suits included the first wet suits, helping with an atmosphere whose dense molecules absorb heat from the body much more rapidly than the atmosphere of air. Gear to speed movement through such an atmosphere was crucial, from swim fins to underwater scooters and sleds that could transport cumbersome film equipment. Further expanding the reach of underwater photography and cinema were more general developments in film technologies. Notably, faster color film opened the way for using 35 rather than 16mm for filming beneath the sea.

The technological history is too intricate to detail here, for engineers and diving teams from nations in the vanguard of industrial modernity, notably with access to coastlines on warm water, were experimenting with the same problems, competing with one another, and devising related, slightly different solutions.[9] The figure credited with inventing scuba, Jacques-Yves Cousteau, for example, was competing with Austrian Hans Hass, although Cousteau never mentioned Hass, who is less well known today in the Anglophone world.[10] Hass too was pioneering scuba, inventing a rebreather using chemicals to scrub CO_2 from the air in contrast to the filled compressed-air tanks used by Cousteau.[11] Hass also was inventing underwater technologies for photography and film. He developed, for example, the water-housing for the Rolleiflex, the Rolleimarin housing, the gold standard for professional photographers for years after it was patented in 1953. Cousteau offers a good example of how the divers and engineers driving such expansion both pursued national interests and came together in international teams.[12] He collaborated with the legendary MIT electrical engineering professor Harold Edgerton in developing underwater lighting, working with the Woods Hole Oceanographic Institution in Massachusetts and subsequently the Scripps Institution of Oceanography in La Jolla, California.

These innovations afforded access to a great variety of marine ecosystems and a range of depths, as well as ways to portray them with greater clarity, with new kinds of movements, from different angles, and in mesmerizing color.[13] Such new capacity was the precondition for the revolution in underwater filming, but these possibilities then opened a new world of aesthetic as well as technical problems. Work was exceptionally difficult in the toxic atmosphere of the undersea. Along with figuring out how to manage these conditions, filmmakers had to display this novel environment so it would be legible as well as gripping to audiences with no prior exposure to it. The challenge was all the greater because the undersea that could now be exhibited differed greatly from the fantastic representations of it through the centuries. Scientists, engineers, and explorers really were plumbing from their coastlines an alien world repeatedly compared to outer space, the other cosmic frontier of the post–World War II era. The distinguished French director Louis Malle started working for Cousteau while a young student in film school, which he left to film *The Silent World* (1956). Because of his contribution, Malle was promoted from cameraman to codirector by Cousteau when they were editing their footage. When Malle looked back on making this film from the perspective of his entire career, he commented, "We had to invent the rules—there were not references; it was too new."[14] A similar sense of creating without precedent was expressed by those involved in Disney's *20,000 Leagues under the Sea*. "Everything they were doing down there was experimental. It was a pretty amazing operation," said Roy E. Disney, Walt's nephew and senior Disney executive at the time.[15]

I focus on just one problem filmmakers faced in creating entertainment from their new undersea access: how to organize this environment into a dramatic setting capable of holding characters, animals, and action. Creating a setting is of course necessary for any representation of place in film. The marine environment presents unique obstacles due to the obscurity of water as an atmosphere, which leads to optical perception differing dramatically from optics on land. To record on color film in such darkness depended on innovations like strobe lighting; however, there was no way to overcome foundational physical limitations, notably the condition that the farthest the eye can see even in distilled water is only about 250 feet. As I explain elsewhere, the obscurity of underwater optics makes the underwater atmosphere one that defies foundational Western conventions for rendering landscape, dating back to the Renaissance.[16] These landscape conventions, such as linear perspective, were so influential in the West that they were taken up in conventional cinema. However, achieving the deep vanishing point that organizes linear perspective was a challenge in an obscure environment that, moreover, was always in movement. "How can life be organized in a world without horizons?" asked Philippe Diolé, a writer who dove with Cousteau's French Navy team in *The Undersea Adventure* (1951).[17]

Diolé was noting the diver's disorientation, which was a safety issue and also provoked his consideration of the unique phenomenology of perception in the submarine environment. Spectators with no underwater experience might be confused or put off by this alien phenomenology, before they had comprehended its interest. Filmmakers needed to show the hazy, disorienting properties of optics underwater in a way that would produce a sense of familiarity and mastery important for emotional involvement. In his reflections on *The Silent World*, Malle recalled Cousteau's taste for what Malle denigrated as "conventional spectacle," "resorting to the techniques of what would be called today docudrama."[18] Malle, in contrast, recounted his youthful enthusiasm for austere documentary authenticity, the imperative "to stay absolutely true," which he absorbed from the sparse, elegant style of the director Robert Bresson. While respecting Malle's sober vision, we can grasp why the bizarre (by terrestrial standards) features of underwater perception might lead Cousteau to turn to recognizable conventions, even clichés. In depths never before glimpsed, let alone exhibited, where audiences had no reference points, the appeal to a preexisting imaginative repertoire was helpful in easing spectators into this strange new world.

To show how filmmakers oriented viewers in the depths, I use examples from two of the most influential films of the era, so popular that they qualify as spectacles, seen by millions of people around the globe.[19] *20,000 Leagues under the Sea* (1954) was the Walt Disney studio's first live-action film, directed by Richard Fleischer, using Technicolor, and essaying the wide screen format of Cinemascope. *20,000 Leagues under the Sea* was nominated for three Academy Awards—Best Art Direction—Color; Best Special Effects; and Best Film Editing—and received the first two. The battle with the giant squid was singled out, but the artfully staged underwater scenes presumably played a role as well.[20] *20,000 Leagues under the Sea* was the second-highest-grossing film of the year in the United States (after *White Christmas*). The publicity for the underwater sequences in the film had piqued audiences' interest even before the film appeared in December 1954. In February 1954 *Life* ran a photo-essay, "A Weird New Film World: Jules Verne Movie Produces 20,000 Headaches under the Sea," showing how those sequences were filmed.[21] "Hollywood's Verne" was the subject of a *New York Times* article dated

March 28, 1954, also explaining how the film was made.[22] To stimulate interest in the film's release in Great Britain the following year, the British cinema fan magazine *Picturegoer* ran its version of the article: "20,000 Headaches."[23]

The second film, Cousteau and Malle's *The Silent World* (*Le monde du silence*, 1956), was still more celebrated on both sides of the Atlantic for its artistic innovations. In 1956 *The Silent World* won the highest honor at the Cannes Film Festival, the Palme d'Or (and is the only full-length documentary to do so). It also won an Academy Award in 1957 for best foreign film. The *New York Times* film critic Bosley Crowther, whose reviews could be cutting and derisive, called it "the most beautiful and fascinating documentary of its sort ever filmed."[24] When the influential, usually trenchant André Bazin reviewed *The Silent World* in 1956, he opened by denigrating the power of writing to capture how captivating the film was. He declared that there was a "ridiculous" aspect to undertaking criticism of this work, as "the beauties of the film are first of all those of nature, and one might as well criticize God."[25] What Bazin downplayed—and perhaps was not privy to, given the unfamiliarity of processes for filming underwater at the time—was how its "beauties" due to nature had been created through artistic use of technological innovation.

Disney's *20,000 Leagues under the Sea*: Setting a Theatrical Stage

The "monster" disrupting nineteenth-century global shipping is not a creature but a fantastic submarine. Verne's Professor Aronnax, played by Paul Lukas, and his assistant Conseil, given a sinister cast by Peter Lorre, make this discovery when they survive an attack on the US warship conveying them to investigate the mystery. Stranded at sea, they clamber onto a platform that leads into the navigation room of a Victorian-era steampunk craft, filled with gaudy instruments of measurement. Its atmosphere of nautical precision contrasts with the luxurious parlor Aronnax next steps into that is bathed in a flickering blue-violet, almost otherworldly, light. The light turns out to emanate from a punched-out picture window, which is an enormous porthole into the depths of the sea. A reverse shot from the outside shows a few green fish gliding by, but when the film gives spectators the vantage point of Aronnax and Conseil, we do not see an aquarium's expected denizens. Rather, Aronnax and Conseil gaze down at a procession of divers on the seafloor, clad in what at first glance appear to be helmet diving suits—only their breathing apparatus is self-contained.

In the following entrancing sequence, less than two minutes long, we witness the divers moving in double file across the screen, carrying on a bier a prone humanoid form draped in white (fig. 1). The divers' movements are grave, in keeping with the physical difficulty of moving through water and the mood of the scene is enhanced by mournful music. At the head of the procession is a figure bearing a cross encrusted with shells and baroque undersea creatures. A brief ceremony follows; however, since the science fiction fantasy has not given microphones to the divers, the leader conducts the ceremony with exaggerated gestures. The leader raises his hand in benediction, and divers kneel in front of a bier before they slowly process back toward the submarine. Then the leader stops and with another

Figure 1. Burial procession in Richard Fleischer, *20,000 Leagues under the Sea*, 1954.

melodramatic pose points toward Aronnax and Conseil, watching from the *Nautilus*. With a sweeping wave of his arm, the leader urges his divers on. The scene shifts to the surface, where Aronnax, Conseil, and harpooner Ned Land, their companion in the novel as in the film, where he is played by Kirk Douglas, are subsequently taken prisoner.

Aronnax's exclamation explains the action: "a burial ceremony, under the sea."[26] The burial procession is the first view into the depths offered by Fleischer in *20,000 Leagues*. From the masterful illusion, one would have no idea of the enormous difficulties shooting the underwater footage subsequently edited into a scene that lasts for a minute and a half.

The first technique for creating plausibility in this strange new environment belongs to the domain of narrative rather than underwater filmmaking. In taking Verne's beloved novel as an inspiration, the film catalyzed the audience's preexisting imaginative repertoire. The voice-over of Professor Aronnax, who is the narrator in the novel, helps us match passages in the most popular playbook of modern underwater fantasy with the physical environment that Verne could only imagine but that the film is able to show. The gravity of the sequence, addressing the most difficult and awe-inspiring rite of passage in human existence, also summons a preexisting mythology of the undersea as a grave reaching back to antiquity. In his review of *The Silent World* two years later, Bazin would emphasize the twinned conjunction of science and myth as a powerful allure of films made underwater. His comments on undersea documentary, "born of science" but revealing "the wonder of the human spirit," apply as well to the underwater sequences of *20,000 Leagues*, which are documentary in their setting beneath the waves—shot off Lyford Cay in the Bahamas—even as they integrate this setting into a fictional narrative.[27]

Along with this powerful appeal to a preexisting imaginative repertoire, the filmmakers ease viewers into underwater space by finding ways to make it resemble in some measure settings on land. In particular, they seek to overcome the shallow depth of field in the underwater environment using the conventions of proscenium theater, a terrestrial setting whose illusionism addresses a similar problem. They translate to the undersea a number of techniques of proscenium theater for creating a sense of deep space within a compressed interior as well as opening enough room for complicated actions among multiple characters.

The first technique is the elevated viewpoint. This angle of viewing, corresponding to the choice box seats in a theater, enhances a sense of depth by giving the eye a greater distance to traverse. The view from above also offers the viewer the feeling of mastering the action, enabling the eye to place elements in relation to one another. *20,000 Leagues* justifies this technique by evoking audience expectations about viewing marine scenes. The viewing window of the *Nautilus* recalls the large viewing windows of aquariums, already popular at the time of the novel's publication (the modern public aquarium debuted in 1853). Indeed, when Verne's first illustrators, Édouard Riou and Alphonse de Neuville, imagined the giant windows of Nemo's *Nautilus*, they modeled them on the aquarium vistas offered by the Paris World's Fair in 1867. The initiation of an underwater spectacle by evoking the frame of the aquarium is hence a second way that the filmmakers appeal to a preexisting imaginative repertoire, which encompassed at the time not just aquariums but also previous ways that cinema showed the undersea. As Crylen and others have observed, J. E. Williamson's photosphere launching underwater filmmaking in 1914 framed the depths as if from the window of an aquarium. What Crylen calls in his title "the cinematic aquarium" dominated underwater vistas in film until the 1940s.

In an aquarium, viewers see a scene set in landscape orientation, or perhaps with a viewpoint tipped slightly upward, if the tanks are on as massive a scale as they were in the 1867 World's Fair. When Fleischer used this view in *20,000 Leagues*, he modified the angle of viewing to tilt it downward at thirty to forty-five degrees (fig. 1). Such an angle enhanced the roominess of the setting revealed, which was actually a narrow space. Its impression of depth also arises from the use of flats to create "overlapping planes," a technique taken from proscenium theater. These planes are "helpful for giving the illusion of depth," since it "is difficult for an audience to gauge the actual depth of the stage behind a proscenium arch."[28] The flats beneath the sea are natural: shapes of coral reefs. The divers' procession from left to right between these flats further puts the viewer at ease, corresponding to the habitual movement of the eye in the Roman alphabet's culture of reading. While the divers look as if they were walking without artifice, the comments by professional diver Bill Strophal, one of the underwater stuntmen for *20,000 Leagues*, speaks to illusionistic power of the staging. In fact, the divers were so tightly organized between the "two rows of coral" that they had difficulty executing the funeral rite in the screenplay. "We were supposed to turn around to the side and pick up the coral and gently place it on the body but the only problem was there wasn't enough room to turn around."[29]

In the theater, techniques to expand the sense of space between the flats also include use of smoke, scrims, and lighting, confusing the eye about the positioning of figures in relation to one another. With coral reefs serving as flats, the film could harness the pervasive haze underwater to similar effect. However, as Fleischer observed, "the sea is there to defeat you."[30] The turbid undersea had to be managed so it could throw a cloud over the action but not obscure it altogether. Among the

Figure 2. Laying hemp carpets in Mark Young, *The Making of "20,000 Leagues under the Sea,"* 2003.

difficulties in filming mentioned in the 1955 *Picturegoer* article, people on the set kicked up silt from the seabed, which became "a fog of sand."[31] Dive supervisor Fred Zendar found a solution (fig. 2). Laying huge hemp carpets kept the sand in place—not for long, with all the people walking, but, as Fleischer observed, enough to obtain several takes.

The burial sequence further drew on the conventions of proscenium theater with its arrangement of bodies. With the floor of the stage serving as a ground plane for creating recession back to a vanishing point, bodies can be placed on a slight diagonal to amplify the impression of receding space. In the underwater burial scene, the figures are arranged diagonally on the seafloor to wrest recession out of a blurry atmosphere. The arrangement along diagonal lines to expand the viewer's sense of deep space is still more pronounced when the *Nautilus*'s crew returns to the submarine (fig. 3). As with the aquarium window, the diagonal arrangement respects Verne's imagination. Verne foresaw divers with self-contained breathing apparatus, which would not be invented for seventy-five years. However, his science fiction divers remained weighted like helmet divers, the only type of diving in his time, walking on the seafloor—in contrast to the fishlike ability of scuba divers to swim in all three dimensions of the underwater atmosphere. The upright stance of helmet divers facilitates their diagonal array in relation to a horizon. The underwater haze aids the effect. It blurs the vanishing point, along with our ability to count the divers, enabling us to imagine a robust crowd and hence fostering considerable depth of field.

20,000 Leagues submerged similar techniques of proscenium theater in the more extended subsequent underwater episode. This episode condenses several dives in Verne's novel, including the hunt in the submarine forest of Crespo and visits to Nemo's underwater cultivations that sustain the novel's renegade collective, which eschews consuming products from land. As in the burial procession, the scene uses coral reefs as flats to confuse the eye about the distance between foreground and background. It anchors the characters on the ground and arranges them at a diagonal. It naturalizes the underwater procession again with movements

Figure 3. Returning to the *Nautilus* in *20,000 Leagues.*

from left to right. The film's focus on a display of Nemo's extensive marine exploitation further affords a better opportunity to show marine life than if the filmmakers had used the hunt in the forest of Crespo, since animals are easier both to film and to view when they are constrained than when they are swimming freely.

The conclusion to this sequence adds another set to the coral reef used in both the burial scene and the harvest scene. This set is the interior of a sunken ship. While Verne's novel does include descriptions of sunken ships, they are viewed from the *Nautilus* and hence offer tableaux of catastrophe rather than open a setting for the characters' adventures.[32] In the Disney film, in contrast, Land and Conseil sneak away from the scene of harvesting and explore the interior of a sunken ship for buried treasure.

The set of the wreckage interior is a stunning way to wrest a space scaled to the human body from the boundlessness of the undersea. In organizing this interior (fig. 4), the film again turns to proscenium theater: in this setting, the conventions of the box set for a room. As in the construction of a room on land, set designers used an architecture of diagonals. In the sunken ship, a parallelogram of the ship's hull tilts to the right, creating a sense of its curvature extending into the depths. The cabin is cluttered with broken-down decor, which, in a fashion similar to the irregular coral reef in the previous scene, gives the eye distances to traverse within a relatively confined space. The broken-down decor exudes mystery appropriate to the environment, rhyming with the visual uncertainty about the depth of field amid the haze. This uncertainty is reinforced by the sea glimpsed through broken timbers or former windows, which prolong vision out into the undefined environment beyond. The diver glimpsed through the rightmost opening in the wreckage in figure 4 looks as if he were part of the work of reef exploitation. In fact, however, had these reefs been close enough to provide such a clear view of the diver, Land and Conseil could not plausibly have slipped away.

These two conspirators have not been investigating the wreck many seconds before the viewer glimpses through the wreckage of a window a shark gliding by in the murky blue, unseen by the characters. If a pistol has been laid on the table, it should be fired, runs the famous axiom by Chekhov. The same is true of sharks in

Figure 4. Interior of a sunken ship in *20,000 Leagues*.

the theater of the sea. The wreck scene ends with Nemo's discovery of the characters' escape and Land's rescue from the shark by Nemo. The episode of the shark is an exception in the adventure episodes of *20,000 Leagues*. In this film, drama primarily occurs in the atmosphere of air, and the undersea is used for novelty and mood. However, the film was made at a time when Verne's fantasy of divers untethered from the surface was fact. The possibilities of scuba could now be the subject of film as well as the means to make it.

The Silent World: Submerging Shot Conventions of Terrestrial Film

The Silent World takes us beneath the surface even before the credits, its cold opening igniting with the burst of an underwater flare. The white-orange light crowning the flare contrasts with the turquoise water even as the contrast highlights its beauty. The opening sequence shows us a group of scuba divers carrying flares, who descend on a diagonal angle with powerful grace and focus over forty-five riveting seconds—the time that the flare lasts—bearing the flares like torches in their muscular arms (fig. 5).[33] The sequence intercuts the diagonal line of the divers' descent with the upward movement of oxygen bubbles from the flares and their regulators. The bubbles are white and their forms are pure and symmetrical, as the medium of water reveals the harmony of the element that sustains life on land, invisible to terrestrial eyes. Strangely, the bubbles swell as they rise, with their bottoms flattened, showing the pressure exerted by the deeper, heavier atmosphere on their ascent.[34] Like the mushroom-capped bubbles, both the flares and the divers' plunge are as mesmerizing as they are provocative. How can fire burn in water? What technologies enable divers to breathe in this irrespirable atmosphere? How can they maintain such control, as their plunge defies the habitual terrestrial experience of vertical descent as falling?

The descent of the divers that opens *The Silent World* uses a different solution to the problem of establishing the viewer's sense of panoramic ease underwater than the burial sequence in the Disney studio film. While Fleischer borrowed techniques from theater, Cousteau and Malle adapted the conventional shots of cinema

Figure 5. The descent of the divers in Jacques-Yves Cousteau and Louis Malle, *The Silent World*, 1956.

to the new possibilities of the underwater environment for perceptual experience and orientation. The descent through vertical space in *The Silent World* submerges a type of establishing shot, which introduces a location where the action will unfold. This type of shot is the crane shot, "commonly used to gradually reveal the ground scale of a location or environment as the camera is moved upward including more details as the camera's vantage point gets higher."[35] To this point, I have emphasized the difficulties posed to filming by the underwater atmosphere, but the filmmakers realized that it also offered aesthetic opportunities. In Malle's words, "The camera by definition—because we were underwater—had a mobility and fluidity; we could do incredibly complicated equivalents of what, on land, would be a combination of crane movements plus enormous tracking shots—and we could do it just like breathing because it was part of the movement of the diver."[36]

The crane shot is a conventional way to achieve the elevating scale of the vertical establishing shot on land. Even as Cousteau and Malle produce a crane shot with scuba, they adapt it to emphasize the vertical axis of the underwater environment. Rather than expanding the viewer's perspective by gradually rising up, the camera expands its field of view as it tracks a descent. The camera remains elevated above the divers, and thus the angle of viewing becomes increasingly steep. In offering this perspective, the opening sequence differs from the crane shot on land, which can take in more of the landscape as it rises and thus expand the viewer's sense of the setting to a distant horizon. However, the intensifying verticality of the sequence, as it follows the divers down, has a power of its own. Steep angles of viewing are dramatic on land, as we experience looking down over the edge of a great height. Cousteau and Malle facilitate the translation by the spectator of the drama of verticality from land to sea by having the divers plunge toward the seafloor, a firm limit resembling the terrestrial ground, rather than letting the scene fade off into the undefined underwater haze.

For Romantic artists such as J. M. W. Turner and Caspar David Friedrich, plunging verticality was a signature of the sublime: the pleasure we take in representations of overwhelming, indeed terrifying, power, as long as we are not ourselves in danger. Thus Turner's watercolor *The Battle of Fort Rock, Val d'Aoust Piedmont,*

Figure 6. J. M. W. Turner, *The Battle of Fort Rock, Val d'Aouste, Piedmont, 1796*, 1815.

1796 (1815), emphasizes the terror of the steep ravine, which is the centerpiece of the image, fringed by the narrow road where the narrative subject of the painting is playing out on the left (fig. 6). The vertical axis achieved at the end of *The Silent World*'s opening sequence does evoke the overwhelming might of the mysterious ocean, yet it creates more enigmatic emotions than vertiginous terror. The divers are descending to formidable depths. Fifty meters down is beyond the reach of the casual recreational diver, even today. They glide, however, with an ease showing mastery, defying the menace of falling that gives sublime images of plunging verticality their impact. Gliding, propelled by swim fins, the divers appear to be flying in a liquid sky. Almost every scuba pioneer, including Cousteau as well as Diolé in *The Undersea Adventure*, likened underwater movement to flying. The comparison is still used by many who try scuba for the first time to express their sense of freedom in the three-dimensional atmosphere of water.[37] Crylen connects *The Silent World*'s sequences celebrating the freedom of scuba with Cousteau's idea that humans would progress to a completely aquatic existence, which Cousteau termed *homo aquaticus* in the early 1960s. Torma too remarks on this fantasy, which she calls the "mermen" (a term in circulation throughout the era of the 1920s–1950s), noting the fascination of both Cousteau and Hass with how scuba enables people to behave like fish.[38]

When Aronnax first witnesses underwater action in *20,000 Leagues*, he explains to audiences the strange scene they behold: "A burial ceremony—under the sea." In *The Silent World* no narration accompanies this initial descent. "At fifty meters beneath the surface, men are making a movie," remarks the narrator of *The*

Silent World following the opening and credits, only explaining the enigma of what we are watching after it has been shown.[39] Yet Cousteau and Malle are similarly appealing to a well-known cultural narrative in this opening sequence, although they do so visually rather than in words. Bazin described the divers' realization of "the old myth of flight" in *The Silent World*, likening them to Icarus in "the admirable painting by Brueghel," although in this painting we see only Icarus's legs as he dives presumably to his death in the water, punished for transgressing human limits.[40]

Cousteau's divers, in contrast, expand the reach of human empire, striking poses that recall neoclassical figures for the advance of civilization, such as the Statue of Liberty holding a torch upright or the torchbearers of the modern Olympic games. Light illuminating the darkness is an emblem of knowledge, and it also is necessary for shooting underwater film. A classical icon of human progress thanks to fire is Prometheus, who stole it for humans from the gods. In *Window in the Sea* Cousteau identified the French marine biologist Louis Boutan, the inventor of underwater photography with "Prometheus [who] had given mankind fire and thus artificial light."[41] In some well-known images, Prometheus takes fire down from Mount Olympus by running, but in others, as in the sketch by Peter Paul Rubens (1636–37), Prometheus descends through the air. Rubens's Prometheus looks over his shoulder furtively, expressing his own trepidation crossing the boundary between humans and gods (fig. 7). There is no residue of this sense of transgression in Cousteau and Malle's Promethean divers descending, although managing the physical dangers of the undersea is a theme in the film.

That Cousteau and Malle are constructing a myth of the diver as modern Prometheus—an example of the film's "docudrama"—is evident if we note the artifice in details of the scene. In most subsequent dives the divers wear wetsuits, which were new technologies at the time. They were necessary even in the comparatively warm seas where the film was shot, given that "bodily heat lost in sea bathing is enormous, placing a grave strain on the central heating plant of the body."[42] Stripped of wetsuits, the divers with their muscular physiques resemble statues. This display of the athletic body also tamps down the posthuman quality of their technologically obscured countenances. Further, the divers' flares evoking the light of progress did not have sufficient power to serve as lighting to film the scene. Twenty years later Cousteau explained that flares were unsuitable for color imagery. In *Window in the Sea* he wrote that "the flare with its thick stream of bubbles is beautiful in itself, but its use as a light source is somewhat limited."[43] Like the fish illustrating this point in *Window in the Sea*, the divers would have "shown up only as . . . dark silhouette[s] had the photographer not used other lighting."[44] That lighting was designed for Cousteau by the MIT electrical engineering professor Harold Edgerton. While critical, this lighting also produced challenges for naturalistic scenery, as it required power from "bothersome cables streaming down from surface vessels or shore" that would "encumber the scene."[45] The iconic figures of *The Silent World*'s opening appealed to the audience's imaginative repertoire by masking aspects of undersea work and of the scene's production.

Figure 7. Peter Paul Rubens, *Prometheus*, 1636–37.

One of the remarkable features of *The Silent World* for publics of the time was the diversity of underwater environment it brought to audiences at home. In each of these environments, from beautiful coral reefs to the greatest depths attained by underwater photographers, the film would face a similar challenge of contending with the strangeness of underwater perception and the limits of underwater optics. In each of these environments, as well, the directors submerged conventions of cinema and adapted them, finding opportunities amid the challenges of submarine materiality. A good comparison with the Promethean descent of the divers is an episode portraying the feeding frenzy of sharks, which Cousteau and Malle were the first to capture in gory color. As throughout the film, the sequences show new technologies pioneered by Cousteau and his team, in this case the technology of the shark cage. However, while the shark cage is enabling, it does not determine the impact of the scene. Rather, this impact derives from how Cousteau and Malle adapted conventions from horror films of shot/reverse shot to show violence from the perspective of both killer and victim.

Figure 8. Shark carnage in *The Silent World.*

The occasion for the feeding frenzy is the carcass of a baby toothed whale, accidentally killed by propellers of the *Calypso,* which has gotten too close to a pod of whales. From the surface, the film shows us disjointed glimpses of menace: the shark's distinctive dorsal fin, chunks taken out of the carcass, and white, churning water stained red, with reflections obscuring the action. In a gesture similar to the opening adaptation of the establishing shot to emphasize verticality, the episode of shark carnage adapts the shot/reverse shot conventions of horror sequences to the underwater environment. Most powerfully, its drama derives from the contrast between the obscurity of the topside view and the violence beneath the surface. Seen underwater, the action is still more terrifying than we could have imagined. The dorsal fins belong to the sharks, whose menace only grows as the spectator sees their full forms and savagery. The chunks turn out to belong to a massive carcass, and the red on the surface emanates from the blood pouring out with each bite (fig. 8).

While the camera underwater is capable of great fluidity when handheld, as we have seen in the opening sequence, in the depths Cousteau and Malle show shark carnage with jerky camera movements, out of focus or dramatically tilted shots, and abrupt cuts. The shot/reverse shot not only contrasts topside and depths but also is used below, showing us both the shark-eye view and views of the shark lunging at the camera, so close that spectators can put themselves in the place of the carcass. Color helps us make out the action amid the confusion: gray and white signify the shark; the whale carcass is black with white flesh. Both are swirled together with the blue of the water, white reflections from the surface, and a cloud of red blood.

As with the divers' Promethean descent, *The Silent World* enhances impact by referencing a preexisting imaginative repertoire—in this case, through narration reiterating the age-old terror of sharks. "The hellish circling of sharks around their prey—if men have seen it, they have not lived to tell the tale," the narrator remarks.[46] At the time of the film, the circling had recently been described by Ernest Hemingway in *The Old Man and the Sea* (1952), a book singled out when Hemingway

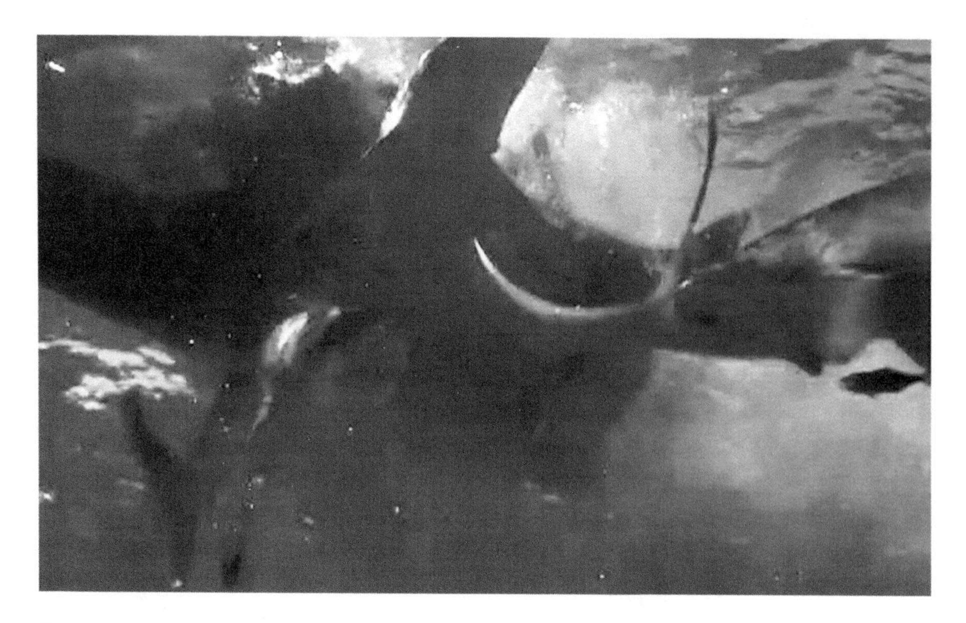

Figure 9. Shark carnage in John Sturges, *The Old Man and the Sea*, 1958.

was awarded the Nobel Prize in 1954. *The Silent World*'s narrator repeats the imaginative glue of age-old terror as the shark sequence unfolds. "Seamen the world over hate sharks," he declares. "For us divers, the shark is the deadly enemy."[47] As in *The Old Man and the Sea*, the seamen react to the feeding frenzy of the sharks with a savage butchery of their own.

In Hemingway's narrative, the fisherman kills the sharks first to try to save his epic catch and then in frustration because he is ineffectual. In *The Silent World* the seamen react to a feeding frenzy they have created with no practical aim, after the *Calypso*'s propellers have killed the baby tooth whale. Indeed, the sharks are engaging in natural behavior, not butchery for its own sake, and today's viewer may wonder whether the horror-film conventions might have more aptly been applied to the crew. Cousteau would subsequently change his stance and demystify the relentlessly bloodthirsty quality of sharks, filming divers working in their presence. At the time of *The Silent World*, he must have been aware of Hans Hass's pathbreaking representations of sharks in *Menschen unter Haien* (*Men among Sharks*, 1947), a black-and-white film Hass made before scuba, instead free diving and using a helmet without a suit in the warmth of tropical waters. In the vanguard of shark documentation, Hass showed the sharks going about their business as the divers film them warily, but without violence, fending them off by aggressive return behavior as necessary.

Nonetheless, Cousteau and Malle's conventions of shark carnage would become iconic across the filmography of shark predation. Just two years after *The Silent World*, the film of *The Old Man and the Sea* (1958) showed the sharks devouring the marlin with the same organization of topside/underwater shots contrasting surface and depth, as well as with clouds of blue, gray, and white tinged with red in

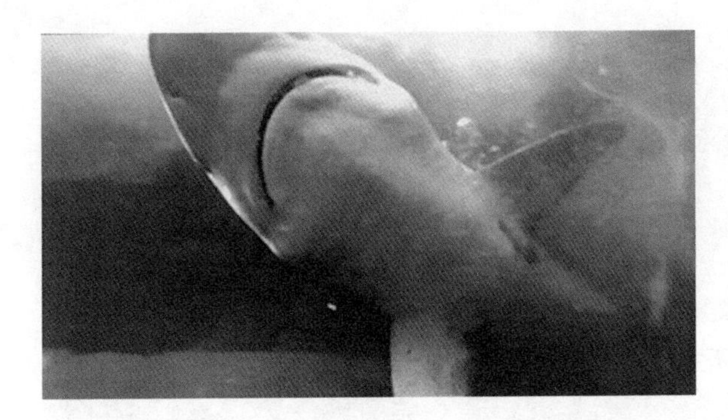

Figure 10. Shark carnage in Peter Gimbel and James Lipscomb, *Blue Water, White Death*, 1971.

the chaos of feeding, and similar shots of sharks roving around the carcass, including one of a shark coming dramatically head-on into the camera (fig. 9). It is hard to imagine that *The Silent World* was not seen by the leading underwater cameraman of the era Lamar Boren, who shot the undersea portion of *The Old Man and the Sea*, as is evident in the comparison of shark carnage from the depths in figures 8 and 9. Peter Gimbel and James Lipscomb's documentary *Blue Water, White Death* (1971), and then Steven Spielberg's blockbuster *Jaws* (1975), inspired by this documentary, expanded these iconic conventions of shark violence into a full-length film. Compare the image from *Blue Water, White Death* in figure 10 to shark carnage in figures 8 and 9. The conventions of shark carnage are still vital today, used notably in summer thrillers such as the sequels to *Jaws* or, more recently, Jaume-Collet Serra's *The Shallows* (2016) and John Turtletaub's *The Meg* (2018). They have gone on to play such an outsized role in how real sharks are imagined that shark conservationists are often filmmakers who have sought to erase these conventions as part of their work. Thus Howard and Michele Hall's IMAX *Island of the Sharks* (1998) depict sharks as majestic, natural inhabitants of the most beautiful oceans of the planet.

I initiated this article with the revolution in underwater technologies that expanded possibilities for both accessing and filming the underwater environment. Across my discussion of *20,000 Leagues under the Sea* and *The Silent World*, I have shown the power of new technologies to enable unprecedented views of this

Figure 11. Spectators in the deep-sea vessel in Wes Andersen, *The Life Aquatic with Steve Zissou*, 2004.

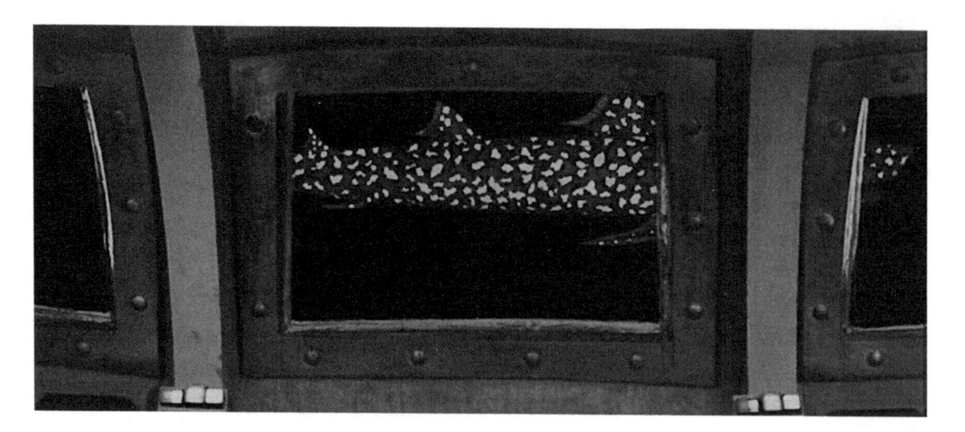

Figure 12. The jaguar shark in *The Life Aquatic*.

environment and also the critical role played by aesthetics in making this environment accessible and entertaining to general audiences. The natural environment is of course always mediated in its exhibition. This mediation is critical in the case of a remote realm few spectators can experience directly that, moreover, defies terrestrial expectations of physics and optics. In the two pioneering works considered in this article, filmmakers made its unfamiliar aspects familiar both by narrowing the gap between the undersea and terrestrial conventions of exhibition and by summoning long-standing imaginative figures for the undersea. This familiarization then enabled filmmakers to showcase the weird, fascinating qualities of this alien realm covering three-quarters of our planet, whose features and inhabitants are still being revealed today.

Further, such a process of adapting topside conventions to organize the depths is a creative contribution to the art of filmmaking. Director Wes Andersen has offered a memorable homage to its history in his *The Life Aquatic with Steve Zissou* (2004). The plot in this film turns around a quest for a mythical killer shark, the jaguar shark, discovered at the film's opening in a topside sequence of shark carnage, using clouds of blood and jerky handheld camera movements—markers of violence unmoored, however, from any glimpse of the animal that is its cause. When the jaguar shark finally appears at the film's end, Anderson creates it with whimsical animation. The shot/reverse shot of shark carnage turns to a shot/reverse shot showing the members of the expedition arrayed in rows in the deep-sea vessel, as if they are spectators at the movies (fig. 11). Staring directly at the film's viewer, they turn out to be entranced by a luminously spotted giant shark, produced through animation, which comes straight at the vessel and then returns to glide by, its massive body broken up by the vessel's windows, as if the shark were gliding across the frames of an analog film strip (fig. 12). To the accompaniment of a melodic, airy, and somewhat melancholy song, "Staralfur," by Sigur Rós, Andersen releases the jaguar shark from documentary, or for that matter from footage shot beneath the sea, to revel in the wonder and artifice of cinematic exhibition.

MARGARET COHEN is Andrew B. Hammond Professor of French Language, Literature and Civilization at Stanford University, where she is jointly appointed in Comparative Literature and English: dlcl.stanford.edu/people/margaret-cohen. Her books include *Profane Illumination: Walter Benjamin and the Paris of Surrealist Revolution* (1993), *The Sentimental Education of the Novel* (1999), and *The Novel and the Sea* (2010). Her article in *ELN* incorporates research for *Underwater Eye*, a book she is completing on the history of underwater cinema. Other projects include coediting *The Aesthetics of the Undersea* (2019).

Notes

1 In Franziska Torma's words, "The act of seeing, as opposed to listening, is crucial to the history of underwater film" ("Frontiers," 28).

2 Rozwadowski, "Arthur C. Clarke," 583. See also Rozwadowski, "Playing by—and on and under—the Sea."

3 Starosielski, "Beyond Fluidity," 150–51.

4 Torma, "Frontiers," 26.

5 Crylen, "Cinematic Aquarium"; Elias, *Coral Empire*. On the history of cinema underwater before scuba, see also Adamowsky, *Mysterious Science of the Seas*. Adamowsky's concluding date is the year of Jacques-Yves Cousteau's first underwater film with scuba, *Sunken Ships* (*Épaves*).

6 Crylen, "Cinematic Aquarium." Sean Cubitt's chapter on *The Blue Planet* in *EcoMedia* focuses on the philosophical significance of the undersea and is less engaged with the technologies of underwater exhibition—with some important exceptions, such as his comments on the lighting system designed by James Cameron to permit viewing of the *Titanic* wreckage.

7 Torma, "Frontiers," 33.

8 On the history of CCOUBA, see Quick, "History."

9 A good overview from the time of some of the camera equipment being developed is Schenck and Kendall, *Underwater Photography*. The acknowledgments do not credit Hans Hass but otherwise distinguish people who played an important role in the history of diving, along with underwater photography and film.

10 The *New York Times* obituary for Hass quotes him commenting that "for Cousteau, there existed only Cousteau. He never acknowledged others or corrected the impression that he wasn't the first in diving or in underwater photography" (comment made to Tim Ecott, dive historian and author of *Neutral Buoyancy*, in Vitello, "Hans Hass").

11 Michael Jung, director of the Hans Hass archive, explains the technologies developed by Hass on this website on rebreathing history: www.therebreathersite.nl/Zuurstofrebreathers /German/hans_hass.htm (accessed June 2018).

12 As a scientific example, the pioneering bathysphere *Trieste*, which reached the deepest point in the ocean for the first time in 1960, was initially used by the French Navy before being sold to the US Navy and was built in Italy according to the design of the Swiss engineer Auguste Piccard.

13 "Only until the recent development of faster color film, high-speed lenses and a wide variety of color correction filters, has the underwater photographer begun to see things in their true colors" (Greenberg, *Underwater Photography*, 37).

14 Malle, *Malle on Malle*, 8.

15 Young, *The Making of 20,000 Leagues under the Sea* (2003), DVD, time code ca. 33:10.

16 See Cohen, "Underwater Optics," 2–3, 7–9.

17 Diolé, *Undersea Adventure*, 23. The rapid translation of this work into English, only two years after it first appeared in French, indicates the wide public interest in discoveries and exploration of the undersea environment. In the same year Cousteau's *The Silent World* was published simultaneously in French and in English.

18 Malle, *Malle on Malle*, 8.

19 The first revolutionary works were documentaries made in wartime by the Austrian Hans Hass (*Underwater Stalking* [*Pirsch unter Wasser*], 1939, and *Men among Sharks* [*Menschen unter Haien*], 1942) and by Jacques-Yves Cousteau (*At Eighteen Meters Depth* [*Par dix-huit mètres de fond*], 1942, and *Sunken Ships* [*Épaves*], 1943). Hass was shooting in Nazi Germany, and Cousteau was a lieutenant in the French Navy in Vichy France. The politics of both pioneers in diving during this period is a subject their biographers have been reluctant to touch, but both have

distanced themselves from the fascist support they received.

20 Compare, for example, the illusionist views in *20,000 Leagues under the Sea* to those in the Technicolor Cinemascope *Beneath the Twelve-Mile Reef* (dir. Robert Webb, 1953).

21 Stackpole, "Weird New Film World," 111.

22 Pryor, "Hollywood's Verne."

23 Hutchinson, "20,000 Headaches."

24 Crowther, "Beautiful Sea," 30.

25 Bazin, "*Le monde du silence*," 35; my translations throughout citations from Bazin.

26 Felton and Verne, *20,000 Leagues*, time code ca. 25:40–42.

27 Bazin, "*Le monde du silence*," 37.

28 Quotations on this page about stage design are from Condee, *Theatrical Space*, 66.

29 Young, *The Making of 20,000 Leagues*, time code ca. 43:30.

30 Young, *The Making of 20,000 Leagues*, time code ca. 32:30.

31 Hutchinson, "20,000 Headaches," 13.

32 Wrecks surveyed in the novel include fragments of La Pérouse's ships *La Boussole* and *L'Astrolabe* (Verne, *Twenty Thousand Leagues*, 130) and a British ship sunk by the *Nautilus* (124), which affords the gruesome vision of corpses in tragic poses. The name of the ship is *The Florida*, an unusual name for a British ship, as Butcher observes, noting that "it must be borrowed from the Confederate *Florida*, which sank 37 Northern ships in just over a year" (408n125).

33 The time the flare lasted was noted by Marden, "Camera under the Sea," 170.

34 Cousteau observed this effect of the bubbles swelling into the lighter atmosphere as they rose in *Silent World*, 6.

35 Mercado, *Filmmaker's Eye*, 167.

36 Malle, *Malle on Malle*, 8. Both Torma and Crylen note the ease of the camera moving underwater. As Crylen observes of "Cousteau's roving undersea images," the liquid atmosphere stabilizes the camera, which "move[s] with a fluidity unusual for handheld camera" ("Cinematic Aquarium," 78, 69).

37 Cousteau: "At night, I had often had visions of flying. Now I flew without wings. (Since that first aqualung flight, I have never had a dream of flying)" (*Silent World*, 6).

38 Crylen, "Cinematic Aquarium," notably his second chapter, "Living in a World without Sun: Jacques Cousteau, *Homo Aquaticus*, and the Dream of Conquering the Deep," 63–106; Torma, "Frontiers," 34.

39 Cousteau and Malle, *Silent World*, time code ca. 2:00.; my translations throughout citations from the film.

40 Bazin, "*Le monde du silence*," 36.

41 Cousteau, *Window in the Sea*, 96.

42 Cousteau, *Silent World*, 12.

43 Cousteau, *Window in the Sea*, 96.

44 Cousteau, *Window in the Sea*, 96.

45 Cousteau, *Window in the Sea*, 60–61.

46 Cousteau and Malle, *Silent World*, time code ca. 53:53. The French is "Le rond infernal des requins autour de leur proie—si les hommes l'ont vu, ils n'ont pas vécu pour le dire."

47 Cousteau and Malle, *Silent World*, time code ca. 56:14. The French is "Tous les marins du monde détestent les requins. Pour nous plongeurs, le requin c'est l'ennemi mortel."

Works Cited

Bazin, André. "Le monde du silence." In *Qu'est-ce que le cinéma?*, 35–40. Paris: Cerf, 2000.

Cohen, Margaret. "Underwater Optics as Symbolic Form." *French Politics, Culture and Society* 32, no. 3 (2014): 1–23.

Condee, William Faricy. *Theatrical Space: A Guide for Directors and Designers*. Lanham, MD: Scarecrow, 2001.

Cousteau, Jacques-Yves. *The Silent World*, directed by Jacques-Yves Cousteau and Louis Malle. FSJYC Production, 1956. DVD.

Cousteau, Jacques-Yves. *Window in the Sea*. Vol. 4 of *The Ocean World of Jacques Cousteau*. Danbury, CT: Danbury Press, 1975.

Cousteau, Jacques-Yves, with Frédéric Dumas. *The Silent World*. New York: Harper and Row, 1953.

Crowther, Bosley. "Screen: Beautiful Sea; 'Silent World' Opens at the Paris Here." *New York Times*, September 25, 1956.

Crylen, Jonathan. "The Cinematic Aquarium." PhD diss., University of Iowa, 2015.

Cubitt, Sean. *EcoMedia*. New York: Rodopi, 2005.

Diolé, Philippe. *The Undersea Adventure*, translated by Alan Ross. London: Sidwick and Jackson, 1953.

Elias, Ann. *Coral Empire: Underwater Oceans, Colonial Tropics, Visual Modernity*. Durham, NC: Duke University Press, 2019.

Felton, Earl, and Jules Verne. *20,000 Leagues under the Sea*, directed by Richard Fleischer. Walt Disney Studios, 1954. DVD.

Greenberg, Jerry. *Underwater Photography Simplified*. 3rd ed. Coral Gables, FL: Seahawk, 1963.

Hutchinson, Tom. "20,000 Headaches." *Picturegoer* (London), July 30, 1955, 12–13.

Malle, Louis, *Malle on Malle*, edited by Philip French. New York: Faber and Faber, 1993.

Marden, Luis. "Camera under the Sea." *National Geographic*, February 1956, 162–200.

Mercado, Gustavo. *The Filmmaker's Eye*. Burlington, MA: Focal, 2011.

Pryor, Thomas M. "Hollywood's Verne." *New York Times*, March 28, 1954.

Quick, Dan. "A History of Closed Circuit Oxygen Underwater Breathing Apparatus." May 1970. archive.rubicon-foundation.org/xmlui/bitstream/handle/123456789/4960/RANSUM_Project_1-70.PDF?sequence=1.

Rozwadowski, Helen. "Arthur C. Clarke and the Limitations of the Ocean as a Frontier." *Environmental History* 17, no. 2 (2012): 578–602.

Rozwadowski, Helen. "Playing by—and on and under—the Sea: The Importance of Play for Knowing the Ocean." In *Knowing Global Environments: New Historical Perspectives on the Field Sciences*, edited by Jeremy Vetter, 162–89. New Brunswick, NJ: Rutgers University Press, 2011.

Schenck, Hilbert, and Henry Kendall. *Underwater Photography*. Cambridge, MD: Cornell Maritime Press, 1954.

Stackpole, Peter, photographer. "A Weird New Film World." *Life*, February 22, 1954, 111–17.

Starosielski, Nicole. "Beyond Fluidity: A Cultural History of Cinema under Water." In *The Ecocinema Reader: Theory and Practice*, edited by Salma Monani, Stephen Rust, and Sean Cubitt, 149–68. New York: Routledge, 2012.

Torma, Franziska. "Frontiers of Visibility: On Diving Mobility in Underwater Films (1920s to 1970s)." *Transfers* 3, no. 2 (2013): 24–26.

Verne, Jules. *Twenty Thousand Leagues under the Sea*, edited and translated by William Butcher. New York: Oxford University Press, 1998.

Vitello, Paul. "Hans Hass, Early Undersea Explorer, Dies at Ninety-Four." *New York Times*, July 7, 2013.

Arctic Nation

HESTER BLUM

Abstract This provocation considers the relationship between the United States and the Arctic. America might be understood as an Arctic nation not just because of its political and resource claims in the polar region but because the nation has become both environmentally and politically inhospitable to human life. The polar regions are no longer climate outliers on the planet, remote regions exceptionally hostile to human life. The United States could do more to recognize forms of geopolitical organization that do not presume continental supremacy; that loose the "territory" from "territorial seas"; that understand the cryosphere as a model for new forms of relation and collaboration; that turn to Indigenous knowledge and traditional ecological knowledge for guidance.
Keywords Arctic, Anthropocene, polar, America, climate change

The elemental fluidity of the seas is both a hydrophysical fact and the first principle for a model of hydro-criticism.[1] Although modern academic disciplines have generally organized themselves around units of analysis such as time periods, nations, data sets, or human societies, an oceanic orientation urges researchers instead to embrace the undulating, nonhuman, nonplanar depths of the sea as a model for critical expansiveness.[2] Oceanic forms of relation do not accede to the signposts or lines of demarcation presumed by territorial spaces; marine, lacustrine, or riparian modes of analysis understand the planet as contingent, solvent, and motile. Even as the sea resists human or terrestrial forms of inscription, however, humans—particularly within the European colonial tradition—have imposed notional cartographic lines to subdivide the globe for navigational and geopolitical ends and to attempt to solidify human positionality in measurable terms. Mishuana Goeman calls this process "colonial spatialization," or the "nationalist discourses that ensconce a social and cultural sphere, stake a claim to people, and territorialize the physical landscape by manufacturing categories and separating land from people."[3] As Goeman's work on Indigenous conceptions of land suggests, geographic forms of inscription do not always track at sea, especially if (following Tim Ingold) the fictive lines of navigation are understood to be a form of writing.[4] If *oceanic studies* designates a field, then *hydro-criticism* designates a practice of disaggregation. A hydro-critical methodology might ask what alternate modes of expres-

57:1, April 2019 DOI 10.1215/00138282-7309688

sion and knowledge projects emerge if the standard were not the linearity of territorial geographies but instead the multidimensional vortices of the aqueous globe. In this sense oceanic studies seeks to move away from Western political demarcations of the globe when studying planetary relations, and toward other analytic dimensions for thinking about surfaces, depths, and the extraterritorial sphere of planetary resources and relations, all of which are suggested by the geophysical, historical, and imaginative properties of the sea. My own work in the field of oceanic studies (most recently on the polar regions) has been anchored by this assumption. The ocean, in other words, exemplifies structures of nonlinear or nonplanar thought. Not a metaphor, the sea has a scalar fluidity that enables the hydrographic world to be at once global and microecological. The modes of inquiry variously known as oceanic studies, hydro-criticism, the new thalassology, and the blue humanities situate the seas and other waterways—not "territory" per se—as central to critical conversations about human and nonhuman relations and exchanges on a planetary scale.[5]

The trace of the human in the sea has a long history, whether in the form of industrial detritus and agricultural runoff or in the still-present atomized bodies of enslaved and jettisoned Africans during the Middle Passage, which, as Christina Sharpe writes, perpetually recycles histories of anti-Black and imperial violence.[6] In our present moment of anthropogenic climate change, human effects are registering in marine environments in accelerating new ways: whether in the volume of microplastics now discernible in Arctic sea ice, or in the Pacific biota now flowing into northern polar waters as the Arctic warms.[7] More perceptibly, states persist in extending sovereignty claims to the sea, as exclusive economic zones at sea spread beyond "territorial waters" and are, in turn, superseded by submarine continental shelf claims, as the United Nations Convention on the Law of the Sea (UNCLOS) permits.[8] Such is the context in which I turn to insistences from a variety of US sources that "America is an Arctic nation."[9] What are the political intentions and critical implications of this assertion when applied to a region characterized by shifting states of hyperborean liquidity? In the meditation that follows, I circle back to the land- and state-based conceptions of the globe proposed by European and American colonialists in order to consider nationalist claims to the aqueous planet. For even as UNCLOS envisions continental borders as subaqueous and porous, it does so in the service of entrenched sovereign claims by nations.

I have heard America's Arctic identity or sovereignty proposed often in the years I have been working in oceanic studies, the polar humanities, and nineteenth-century US literature and culture. The question of what relationship the United States has to the Arctic has been posed directly and indirectly by State Department representatives, meteorology professors, naval lawyers, resources for Arctic and sub-Arctic Native and First Nations holdings at the National Museum of the American Indian, oil and gas barons, Iñupiat and Inuit residents of Alaska, and academic humanists. The range of intentions of these various constituencies in making the case for an Arctic America reflects their diverse interests and investments. For even if hydro-critical scholars do not politicize the ocean, states do. This short essay considers some implications of American claims to the Arctic in our present

Anthropocenic epoch, drawing on a late nineteenth-century US expedition to the Canadian archipelago to throw into relief some of the challenges of importing terrestrial ideologies to oceanic spaces.

The first things that come to mind when US residents think about the Arctic are cold, ice, snow, and polar bears, according to a research survey by the Arctic Studio (2015). A statistically notable number of respondents also indicated penguins, which are not found in the Northern Hemisphere. Survey respondents were asked, "How important is the Arctic to your identity as an American?" Not very, the results demonstrated: "Most Americans ascribed a low importance to the Arctic in relation to their national identity."[10] The spherical distortion that most US-oriented maps exhibit—in which the archipelagic states of Alaska and Hawai'i are displaced and radically out of scale—is one factor in the relative disinterest in and misinformation about the Arctic held by Americans. Brian Russell Roberts and Michelle Stephens have identified such distortions as a reflection of a "continental bias" in the American geographic imagination.[11] The United States and other circum-Arctic nations anchor their interests in the region on the landmasses that each nation claims above the Arctic circle. Yet the concept of "territory" in oceanic spaces is fraught, in part because it privileges the human or the cartographic over the nonhuman world—the ideological over the phenomenal world. As the melting of the polar ice caps causes the seas to rise, the contours of the land that interrupts the aqueous globe are themselves transformed, whether they delineate low-lying islands or coastal cities. Even the relevant legal boundaries are porous. The Law of the Sea stipulates that the "sovereignty of a coastal State extends, beyond its land territory and internal waters and, in the case of an archipelagic State, its archipelagic waters, to an adjacent belt of sea, described as the territorial sea. . . . This sovereignty extends to the air space over the territorial sea as well as to its bed and subsoil."[12] The oxymoronic notion of "territorial waters" serves the strategic purposes of those nation-states making claims to Arctic resources, yet it imposes land-based (and Western) notions of property on a medium resistant to such inscriptions. As the Inuk author Rosemarie Kuptana writes forcefully, "The Inuit Sea is once again discussed in Canada and in the global community in the context of sovereignty and security and in the absence of Inuit."[13] Kuptana's objection calls to mind an analogous history in the United States of oxymoronic legal categories created in the erasure of the voices and the rights of Indigenous people, African Americans, and other people of color: most notoriously, the Supreme Court established Native tribes as "domestic dependent nations" in *Cherokee Nation v. Georgia* (1831), enshrined segregation in the "separate but equal" ruling in *Plessy v. Ferguson* (1898), and judged US island territories "foreign in a domestic sense" in the Insular Cases of the early twentieth century. Understood within this juridical, political, and discursive history, the notion of "territorial waters" decenters and minoritizes the oceanic. In specifying the "bed and subsoil" as included in territorial seas, too, the UNCLOS definition preserves for states their claims to mineral resources for potential extraction.

From a US-oriented geopolitical perspective, Arctic America came into being with the acquisition of Alaska in 1867. In the intervening 150 years the region has been important for US military strategy and shipping security and, increasingly, as a repository of fossil fuels and other natural resources. For these reasons, and despite the Arctic Studio survey's conclusion that US citizens do not self-identify with the Arctic, the polar region remains a hot topic. In the words of the State Department's special representative for the Arctic during the Obama administration, Admiral Robert J. Papp Jr., "The future of America is inextricably linked to the future of the Arctic."[14] The other seven circum-Arctic nations—Canada, Denmark, Finland, Iceland, Norway, Russia, and Sweden—identify similar national value in the North. Arctic nations are attentive, as well, to the effects of climate change on Indigenous Far North residents and on the planet more broadly, especially as the Arctic is a climate multiplier. (A 2°C global rise in temperature would actually result in an increase of 3.5–5°C in the Arctic.)[15] The accelerating melting of the North's polar ice caps and consequent sea level rise, too, is an Arctic crisis with global effects. "From a national security perspective," the former head of a US Navy commission on climate change has argued, "climate change is all about the water: where it is or isn't, how much or how little there is, how quickly it changes from one state (e.g., solid ice to liquid water) to another."[16] If it takes a sense of threat to American military or energy "security" to compel the global laggard United States to respond to global warming (as the sole nation in the world to reject the conditions of the Paris Agreement on climate change), then state-directed interventions in the polar regions may not lack value.

Yet "America" is of course not one nation—not just the United States—but dozens of countries across two continental landmasses. Like the Arctic, the idea of America is subject to misidentifications and territorializations that overwrite what was once fluid (and bicontinental) as a circumscribed singularity. "America" is a land grab. The US government's sense of America as an Arctic nation is dependent on an extension of a continental logic: that the state of Alaska is a territorial foothold granting access to fossil fuel extraction and Northern sea routes for the polity of the Lower 48.[17] National security as pegged to military, economic, and energy resources is not the only frame with which we might think of the United States as an Arctic nation, of course. Another polar model for thinking of America as an Arctic nation recognizes that human survival requires human-nonhuman collaboration, resource preservation, and ecologically responsive infrastructure. The polar regions are no longer climate outliers on the planet, remote regions exceptionally hostile to human life; as they warm, temperate regions themselves become inhospitable by the same processes. (The developing world registers the most extreme effects of anthropogenic climate change but profits the least from the industrialization that propels it.) The United States could do more to recognize forms of geopolitical organization that do not presume continental supremacy, that loose the nonsensical "territory" from "territorial seas," that understand the cryosphere as a model for new forms of relation, that turn to Indigenous knowledge and traditional ecological knowledge for guidance.

In our contemporary Anthropocenic moment of Arctic and Antarctic polar ice-sheet collapse, human life on earth can feel ephemeral, both because of and despite humans' irreversible impact on global climate and the geological record.

Ecological injustice in the forms of pollution, lack of access to clean water, environmental racism, and food inequality have rendered the American climate extreme and hostile. My provocation is that we extend this argument in a more precipitous direction: America is an Arctic nation now because it has become both environmentally and politically inhospitable to human life. The polar regions are key to thinking about human and nonhuman futurity during the Anthropocene, as there is a direct relationship between irreversible anthropogenic climate change and the Arctic. The global trade interests of early modernity, when the first Northwest Passage expeditions were launched in search of faster routes to Asia, inaugurated, in turn, industrialization's appetite for fossil fuels (and increase in human energy consumption). The oil and gas deposits now targeted for extraction would not be accessible had the carbon usage that necessitates their mining not produced the irreversible warming effects presently melting the polar ice sheets. Humans must attend to the nonhuman processes of accumulated and diminishing ice. US claims to the Arctic today reveal an outmoded yet continuous drive to national sovereignty, a drive that was self-consuming for many polar expeditions and continues to be self-consuming today.

A Hard Case

Consider, by way of example, the life and death of Charles Buck Henry (born Charles Henry Buck), a participant in the US-sponsored Lady Franklin Bay Arctic Expedition (LFBE), led by Adolphus Greely, a Civil War and Indian Wars veteran with no sailing experience.[18] The expedition traveled in 1881–84 to Ellesmere Island, the northernmost island in the Canadian archipelago, to participate in the first International Polar Year, a scientific survey conducted by a number of nations. After resupply ships failed to reach their camp, eighteen of the twenty-five men in the LFBE either starved, died of scurvy, or—in the case of Henry—were executed for theft of food. The hostility to human life experienced by expedition members in the Arctic not only takes the forms customary to polar voyaging in its climate extremity but also reflects the limitations of expeditionary practice organized primarily around territorial claims. These imperatives were at play in the LFBE. The life of Private Henry exemplified American sovereign cruelty, as he practiced violence in many of its most common racist, interpersonal, colonialist, and rhetorical forms in the nineteenth century: he had fought in the Indian Wars; was incarcerated for forgery and theft; murdered a Chinese immigrant in territory illegally occupied by the United States; performed minstrelsy songs in blackface; stole food from his starving crewmates, who carried out the captain's order to execute him for it; and ultimately furnished those crewmates with food in the form of his own corpse. Henry's life was a violation of borders, whether territorial, political, social, or corporeal. His habitual contravention of sovereign boundaries did not change its methodology when relocated from the United States to the Arctic: Henry doubled down on territorial forms of violence rather than find new forms of relation in the hydrographic space of the polar regions. His time with the LFBE illuminates what happens when "America" puts the "nation" in the Arctic and consequently extracts and exploits resources in self-depleting ways.

The primary mission of the LFBE was to establish an international weather station in northeastern Ellesmere Island as part of the first International Polar Year, a collaborative, multinational effort to record Arctic climate data. Participants in the International Polar Year included the circum-Arctic nations as well as the Austro-Hungarian Empire, Germany, France, and the United Kingdom.[19] Yet even as the men were engaged in global scientific objectives, Greely's expedition brings into relief some of the limitations of nationalist rhetoric and practices in marine spaces. The ambit of Henry, the executed food thief, provides a narrative thread. Henry joined the US Army's Seventh Cavalry shortly after its 1876 rout by Lakota and other Plains Indians at the Little Bighorn. The Seventh Cavalry continued to skirmish with Native warriors, and Buck served on the edges of US territorial claims in the 1870s and early 1880s. He was imprisoned for forging a commanding officer's signature to requisition whiskey; following his release or possibly his escape, he made his way to Deadwood, South Dakota (a town illegally established by gold miners in Lakota Territory), where he murdered a Chinese man in a gambling confrontation. He was a large man and a "hard case."[20] Reversing his middle name and surname, Henry then joined the Fifth Cavalry, Greely's own division, and was recommended for service in the Arctic—a further frontier for US territorial exploration. Fresh from the Indian Wars, Henry and the men of the LFBE were assisted by two Inughuit or northern Greenlandic Inuit, called by the crew Jens Edwards and Thorlip Frederik Christiansen, who in the usual practice of white Western Arctic expeditions had been pressed into hunting and translation service (and who both died on the venture). Henry's pre-Arctic biography demonstrates the blurred margins between extralegal US claims in Indigenous territories and state-sponsored exploratory missions.

In their Fort Conger winter quarters Henry coedited the expedition's newspaper, the *Arctic Moon*, which found satiric humor in the history of land grabs and governance claims. Writing from a part of Ellesmere Island named for Henry Grinnell, an American titan of shipping and an Arctic patron, a correspondent to the editor offered himself up for public office: "As Grinnell Land is a reorganized territory of the United States and having a Territorial form of Government, a Delegate from this Territory is to be seated who is to take his seat at the opening of the 47th session of Congress." The self-nominee's platform included "liberal appropriation for the purchase of lime-juice, compulsory education, unlimited emigration, a hundred and sixty acres of land, one musk-ox and two Esquimaux dogs to each actual settler."[21] Grinnell Land was not then and never has been a US territory, and the mock candidate's platform invokes a range of contemporary US political issues regarding land use, sovereignty, immigration, and settlement in its references to Western territorial expansion and the forty acres and a mule offered to formerly enslaved Black Americans (albeit in Arctic-appropriate terms). The article's invocation of "a Territorial form of Government" is most directly a reference to the status of territorial claims under American imperialism, but the subsequent joke about trying to work frozen land with a musk ox and two huskies shows that a US sense of "territory" would not mean much in the Arctic cryosphere. This candidate's statement shows the men of the LFBE playing with the expectations of US territorial and post-Reconstruction practices in the few years before America would launch its most aggressive and widespread practices of seizing overseas territories in the Philippines, Hawai'i, and Guam (all themselves archipelagoes).

The crew engaged in other forms of playacting as well, particularly in their Christmas variety show. The presentation included Henry's minstrel performance of "plantation melodies"; another sailor dressed as an "Eskimo belle."[22] One act that Greely recorded in his voyage narrative "was a representation of an Indian council, which ended with a war-dance. . . . Most of the actors had served in the far West, and some had spent months continuously in Indian camps, and so were thoroughly familiar with the parts portrayed. I doubt very much if a more realistic representation of the wild red-man was ever presented in the Arctic Circle, if elsewhere."[23] While these were common forms of "humor" in the nineteenth century—designed to emphasize and mock racial and ethnic differences and to enforce white supremacy through cultural appropriation—the resonance of such performances in the Arctic is less clear. The Native interlude, for one, stages a scene of sovereignty: an Indian council. For the white members of the LFBE, the council becomes an occasion for diversion, a rejection of the legitimization of any form of Indigenous sovereignty. Just two weeks before this performance, in fact, the Inuk hunter Edwards had tried to escape from the expedition and head north. He was tracked by a large party and, according to Greely, "returned to the station without objection, and in time recovered his spirits. No cause for his action in this respect could be ascertained other than his intense desire to return to his home."[24] Edwards's displacement from his people and his acquiescence to his continued custody in the hands of the US voyagers take on particular acuteness when considered as a preamble to the supremacist dominance staged a fortnight later. The sense of *home* experienced by Edwards is insufficient cause, to Greely, for him to claim it. He eventually drowned while hunting for food for the other men.

Later in the expedition, when theatricals had yielded to the spectacle of looming starvation, Greely presented a series of lectures to keep the men occupied. His first talk, nearly two hours long, was on "the physical geography and the resources of the United States," a topic of wish fulfillment in a region that the expedition regarded as barren.[25] Despite the hunting efforts of the Inughuit men Edwards and Christiansen, the absence of material for sustenance became calamitous. Henry was exposed as a serial thief of the expedition's scant food, which included tiny shrimp and *tripe de roche*, a rock lichen that earlier Arctic expedition members (on similarly disastrous missions) had resorted to eating, along with their leather boots.[26] Henry raided their old winter quarters for sealskins and sealskin thongs and was caught "stealing shrimps out of the general mess-pot."[27] After the order to execute Henry had been carried out, Greely noted that "fully twelve pounds of seal-skin were found cached among his effects."[28] Sealskin is minimally edible for a human in extremity, and a particular challenge for one with teeth weakened by scurvy—a disease that could have been avoided had white expedition members been willing earlier, as Inuit commonly were, to eat raw seal, an antiscorbutic. (In this context the *Arctic Moon* proposal for a political platform founded on "liberal appropriation for the purchase of lime-juice," Anglo-Americans' preferred antiscorbutic, takes on special irony.) The LFBE failed to use local resources in communal ways, maintaining notions of privacy and property that were unsustainable in extreme climates. We see this in a diary entry from another expedition member, Roderick Schneider, who was also suspected of (but not caught) stealing food: "Although Henry has told before his death that I had eaten a lot of sealskin, yet, although I am a dying man, I deny the assertion; I only ate my own boots and a part of an old pair of pants."[29] Resources are shared in the extreme climate of the Arctic, as the work done on behalf of the expedition by Inughuit hunters demonstrates. Schneider's defense was that he ate only his *own* boots and clothing: access to food and self-consumption were understood by him as principles of personal ownership, not of the collective good.

After the return to the United States of fewer than a third of the men, when whispers of cannibalism had begun to circulate (and would severely undercut any tragic heroism attributed to the expedition), Greely told newspapers that he knew "nothing of the cannibalism, but that if it was practiced the men did so privately and on their own responsibility."[30] Greely's sense of a "private" form of cannibalism is striking, especially in light of the communal practices that typically govern resource consumption in the Arctic. Individualist claims to resources neglect the collective good and hollow out possibilities for sustainability. What is more, the public shame of private cannibalism propelled the survivors to engage in a bizarre fiction. "To represent Henry's devoured body," newspapers reported, "sticks were tied to the bones and a wooden ball adjusted to the skeleton for a head, and the whole frame wrapped in muslin. This was put into a casket and palmed off for the body of the dead man."[31] The LFBE began as an international collaboration to gather scientific information on the nonhuman environment; it ended in a ghastly show of self-consuming resource mismanagement.

In its outlandish horror, Henry's career in contravention while engaged in US national expansion shows some of the flaws in importing a proprietary Western nationalism to an oceanic space. The reification of unequal terms such as *territorial seas* (like *domestic dependent nations*, or island territories judged "foreign in a domestic sense") claims affiliation for the sake of control, without incorporation or justice. In a continuing Anthropocenic moment of resource extraction, a hydro-critical approach understands Henry's grim career as an argument against legislating visible and submarine boundaries, instead embracing the interdependence and interrelation of the human and the nonhuman, the terrestrial and the aqueous. Attending to such misrecognitions of planetary relation is one imperative of hydro-critical practices. As regions simultaneously fluid and terrestrial, inhabited and not, stateless and multiply contested, the Arctic and Antarctica provide models and resources for nonlinear understandings of movement in time and space. This circular logic brings the polar regions more directly into the sphere of planetary imaginaries, in which we reorient our perspective away from land-based and Western Hemisphere–based visualizations (in which the Arctic and Antarctica are both remote and subject to territorial claims) and toward a centering of what has too long been imagined as the ends of the earth.

HESTER BLUM is associate professor of English at Pennsylvania State University. She is author of *The View from the Masthead: Maritime Imagination and Antebellum American Sea Narratives* (2008), which received the John Gardner Maritime Research Award. Her edited volumes include *Horrors of Slavery* (2008), William Ray's 1808 Barbary captivity narrative; the essay collection *Turns of Event: American Literary Studies in Motion* (2016); and a special issue of *Atlantic Studies* on oceanic studies. Her latest book is *The News at the Ends of the Earth: The Print Culture of Polar Exploration* (2019).

Notes

1 I use the term *hydrophysical* instead of *geophysical* (or related terms like *geographic*) to resist the presumption of "earth" or "land" or "ground" inherent in the *geo-* prefix.

2 I expand more fully on this position in Blum, "Prospect of Oceanic Studies."

3 Goeman, "(Re)mapping Indigenous Presence," 296.

4 In his mediation on the "anthropology of the line," Ingold argues that "any history of writing must be part of a more comprehensive history of notation" (*Lines*, 11).

5 See Blum, "Introduction"; and Blum, "Prospect of Oceanic Studies."

6 Sharpe, *In the Wake*, 40–41.

7 Peeken et al., "Arctic Sea Ice"; Katz, "Alien Waters."

8 *United Nations Convention on the Law of the* Sea, Art. 76, Pt. IV (December 10, 1982), 53.

9 On the US political front, see most trenchantly Papp, "America Is an Arctic Nation." Papp was the special representative for the Arctic in Obama's State Department; visitors to this site in the post-Obama era will find a note at the top of the webpage that reads: "This is historical material 'frozen in time.' The website is no longer updated." All of the Obama White House archives have this notice—the "frozen in time" designation is not a special Arctic pun. Yet global warming has made obsolete the notion that freezing is a state of permanence anywhere on the globe, whether in the form of Siberian permafrost or the Svalbard Global Seed Vault.

10 "Most Americans ascribed a low importance to the Arctic in relation to their national identity," the survey concluded. See Hamilla, "Arctic in U.S. National Identity."

11 Roberts and Stephens, "Introduction," 12.

12 *United Nations Convention on the Law of the Sea*, 27. The porousness encompasses the atmosphere as well: "This sovereignty extends to the air space over the territorial sea as well as to its bed and subsoil."

13 Kuptana, "Inuit Sea," 10. See also the astute discussion of this issue by Waller in "Connecting Atlantic and Pacific."

14 Papp, "America Is an Arctic Nation."

15 Schlosser et al., "A 5°C Arctic in a 2°C World."

16 Titley, "Global Warming a Threat to National Security."

17 This is a colonial logic that I have not generally found of interest or relevance in my own work on polar writing and ecomedia, but, given the quickening of such rhetoric in contemporary discourses on climate change and resource extraction, I explore its implications in this meditation. See Blum, *News at the Ends of the Earth*.

18 The biographical information on Henry is drawn from Stein, "Arctic Execution"; Copley, "Measure of Human Grit"; *New York Times*, "Private Henry's Record"; and Greely, *Three Years of Arctic Service*.

19 Subsequent IPYs were observed in 1932–33, 1957–58, and 2007–8.

20 Henry's size was notable; even after months of privation during the LFBE, he weighed 203 pounds, among a crew whose average weight was 176 pounds. See Greely, *Three Years of Arctic Service*, 252. All future references to this volume are taken from this edition.

21 "To the Editor," *Arctic Moon* 1, no. 2 (1881), Adolphus Greely Papers, 1876–1973, Stefansson Collection (hereafter cited as Steff MSS), Dartmouth College.

22 Greely Papers, Steff MSS 64, box 2: 20, 1. For more on blackface performance in the polar regions, see Mossakowski, "'Sailors Dearly Love to Make Up.'"

23 Greely, *Three Years of Arctic Service*, 131–32.

24 Greely, *Three Years of Arctic Service*, 123.

25 Greely, *Three Years of Arctic Service*, 580.

26 After his first Arctic expedition the British polar explorer John Franklin became known as the "man who ate his boots" during a terrible overland crossing to the Coppermine River delta.

27 Greely, *Three Years of Arctic Service*, 699.

28 Greely, *Three Years of Arctic Service*, 700.

29 Greely, *Three Years of Arctic Service*, 703–4. In a striking incidence of oceanic textual circulation, Greely had access to Schneider's diary through a quirk of North American waterways. As Greely writes in a note, "Schneider's diary, stolen without doubt by a seaman of the relief squadron, was found in a mutilated condition on the banks of the Mississippi River, and was sent to me by Mr. J. A. Ockerson, U.S. Civil Engineer, as these sheets were going to press" (703–4).

30 *Reading Eagle*, "Wooden Man Buried." A similar account appeared in several other newspapers, including *New York Times*, "Victims of a Blunder."

31 *Reading Eagle*, "Wooden Man Buried."

Works Cited

Blum, Hester. "Introduction: Oceanic Studies." *Atlantic Studies* 10, no. 2 (2013): 151–55.

Blum, Hester. *The News at the Ends of the Earth: The Print Culture of Polar Exploration*. Durham, NC: Duke University Press, 2019.

Blum, Hester. "The Prospect of Oceanic Studies." *PMLA* 125, no. 3 (2010): 770–79.

Copley, Frank Barkley. "The Measure of Human Grit: A Traitor's Death in the Arctic." *American Magazine* 71, no. 3 (1911): 330–39.

Goeman, Mishuana. "(Re)mapping Indigenous Presence on the Land in Native Women's Literature." *American Quarterly* 60, no. 2 (2008): 295–302.

Greely, Adolphus W. *Three Years of Arctic Service: An Account of the Lady Franklin Bay Expedition of 1881–1884 and the Attainment of the Farthest North*. New York, 1886.

Hamilla, Zachary D. "The Arctic in U.S. National Identity (2015)." *Arctic Studio*, December 19, 2017. www.arcticstudio.org/ArcticStudio_ArcticInUSNatlIdentity2015_20171219.pdf.

Ingold, Tim. *Lines: A Brief History*. London: Routledge, 2016.

Katz, Cheryl. "Alien Waters: Neighboring Seas Are Flowing into a Warming Arctic Ocean." *Yale Environment 360*, May 10, 2018. e360.yale.edu/features/alien-waters-neighboring-seas-are-flowing-into-a-warming-arctic-ocean.

Kuptana, Rosemarie. "The Inuit Sea." In *Nilliajut: Inuit Perspectives on Security, Patriotism, and Sovereignty*, edited by Scot Nickels, Karen Kelley, Carrie Grable, Martin Lougheed, and James Kuptana, 10–13. Ottawa: Inuit Tapiriit Kanatami, 2013.

Mossakowski, Tomasz Filip. "'The Sailors Dearly Love to Make Up': Cross-Dressing and Blackface during Polar Exploration." PhD diss., King's College London, 2014.

New York Times. "Private Henry's Record." September 12, 1884.

New York Times. "The Victims of a Blunder: More Light on the Dreadful Story of Greely's Camp." August 14, 1884.

Papp, Robert J., Jr. "America Is an Arctic Nation." December 2, 2014. obamawhitehouse.archives.gov/blog/2014/12/02/america-arctic-nation.

Peeken, Ilka, Sebastian Primpke, Birte Beyer, Julia Gütermann, Christian Katlein, Thomas Krumpen, Melanie Bergmann, Laura Hehemann, and Gunnar Gerdts. "Arctic Sea Ice Is an Important Temporal Sink and Means of Transport for Microplastic." *Nature Communications* 9 (2018): 1–12. www.nature.com/articles/s41467-018-03825-5.

Reading Eagle. "A Wooden Man Buried: Why and How Private Henry Was Shot." August 14, 1884.

Roberts, Brian Russell, and Michelle Ann Stephens. "Introduction." In *Archipelagic American Studies*, edited by Brian Russell Roberts and Michelle Ann Stephens, 1–54. Durham, NC: Duke University Press, 2017.

Schlosser, Peter, Stephanie L. Pfirman, Rafe Pomerance, Margaret Williams, Brad Ack, Phil Duffy, Hajo Eicken, Mojib Latif, Maribeth Murray, and Doug Wallace. "A 5°C Arctic in a 2°C World: Challenges and Recommendations for Immediate Action." Briefing Paper for Arctic Science Ministerial, Columbia Climate Center, Columbia University, New York, NY, September 28, 2016. academiccommons.columbia.edu/doi/10.7916/D8640WKN.

Sharpe, Christina. *In the Wake: On Blackness and Being*. Durham, NC: Duke University Press, 2016.

Stein, Glenn M. "An Arctic Execution: Private Charles B. Henry of the United States Lady Franklin Bay Expedition 1881–1884." *Arctic* 64, no. 4 (2011): 399–412.

Titley, David. "Global Warming a Threat to National Security." *Cognoscenti*, February 20, 2013. www.wbur.org/cognoscenti/2013/02/20/climate-national-security-david-titley.

Waller, Nicole. "Connecting Atlantic and Pacific: Theorizing the Arctic." *Atlantic Studies* 15, no. 2 (2018): 256–78.

Ice Thieves

Urban Water, Climate Justice, and the Humor of Incongruity in Jane Rawson's *A Wrong Turn at the Office of Unmade Lists*

TERESA SHEWRY

Abstract Navigating humor's potentials for violence and for creative and critical connections with oceanic catastrophes, from extinctions to sea-level rise, this essay argues that humor is a diverse and perhaps important dimension of contemporary cultural production about the changing oceans. Jane Rawson's *A Wrong Turn at the Office of Unmade Lists* uses humor to emphasize fictional people's challenging and idiosyncratic daily struggles to survive in a future marked by sea-level rise and other climate-related processes. The novel's absurd tragicomic tensions situate climate change as violence that is, at some level, unrepresentable. Its humor stresses not only the immense difficulties faced by characters who navigate in this future but also the deep socioeconomic inequality with which such difficulties are bound up, opening the narrative toward efforts to achieve climate justice.

Keywords sea-level rise, climate change, humor, climate justice, Australia

As Earth's climate changes and people struggle to survive flooding and displacement associated with rising seas, bureaucrats administer offices dedicated to Odd Socks, Partially Used Pens, Unmade Lists, Shadow Storage, and more. These and other characters' whimsical approaches to living amid the violence of a changing climate establish a humorous aesthetic in the Australian writer Jane Rawson's futuristic novel of Melbourne, *A Wrong Turn at the Office of Unmade Lists* (2013).

There seems to be nothing to laugh about in relation to projected futures for Melbourne in the context of climate change, especially when we consider the intersections between the climate and urban water conditions. Home to more than 4.5 million people, the city and surrounding areas are expected to see more extreme heat waves, drought, bushfires, rainfall, flooding, and storms and a decrease in total snowfall and rain. Like most large cities, Melbourne is interwoven with a complex coastal environment. A sea-level rise of two meters—a figure well within the National Atmospheric and Oceanic Administration's (NOAA) range of global

57:1, April 2019 DOI 10.1215/00138282-7309699

mean sea-level-rise projections for the end of the twenty-first century in a business-as-usual emissions scenario—would subject areas of the city to tidal flooding, including parts of Footscray, an important neighborhood in *A Wrong Turn at the Office of Unmade Lists.*[1]

Rawson's novel and other Australian literary works complicate the typically dead-serious tone of narratives about climate change, offering a creative and disturbing art of laughter. One of the most ancient and enduring theories of humor suggests that people laugh at incongruity, including what appears to be disproportionate, absurd, or contradictory. When humor navigates the incongruities exposed in climate-justice narratives—among them people's radically disproportionate contributions and exposures to the violence of sea-level rise and to precarious freshwater conditions—it runs a troubling range of cultural agency. In 2015, for example, the Australian politician Peter Dutton was caught on film joking with Tony Abbott, the prime minister, about the impact of rising sea levels in the Pacific Islands.[2] When the footage was made public, Bill Shorten, the Australian opposition leader, responded to the joke: "The fact that the Prime Minister is laughing along with it reminds me of what Barack Obama said: any leader who doesn't take climate change seriously is not fit to lead."[3]

Such a critique may seem obvious and necessary, but to simply reject humor in relation to climate change would mean erasing the differences in traditions of humor and the possibilities of creativity and critique that humor can bear. No situation is inevitably humorous, of course, in the sense of provoking a physical and emotional shift such as laughter; literary humor gains life through the perspectives brought by its readers. In this essay I read for the humorous potentials of contemporary Australian cultural production about climate change, focusing particularly on *A Wrong Turn at the Office of Unmade Lists,* a narrative that examines the relationships between climate change and urban water. From precarious freshwater resources to sea-level rise, Rawson situates water as an important urban material. She offers a humor of incongruity in a variety of modes, including disproportion, irony, and absurdism. Humorous techniques of disproportion, which examine and trouble dimensional relationships, and of cosmic irony, which emphasize "a contradictory outcome of events as if in mockery of the promise and fitness of things," illuminate fictional people's challenging and idiosyncratic daily struggles to survive, as well as the socioeconomic inequality with which such struggles are bound, opening the narrative toward efforts to achieve climate justice.[4] The novel's absurd tragicomic tensions situate current processes of climate change as violence that is unnecessary and, at some level, unrepresentable. Such humor is disturbing because it thrives on disruptions and disproportions, yet it may be valuable for hydro-critics who seek to complicate common imaginaries of water in terms of flow and connection, for it attends to the ways in which urban water is entangled in experiences of constraint, inequity, and disconnection.

Urban Water

Melbourne is a city built in the ocean and on swamps, rivers, and lagoons. On an 1803 map the English surveyor Charles Grimes refers to West Melbourne, at the time a wetland and saltwater lagoon, as "Swamp."[5] Jennifer Robertson describes the swamps as an important source of food and culture for local Indigenous peoples and traces the devaluation of these ecosystems in a context of colonialism.[6] Settlers drained swamps, filled and channeled expanses of river, dredged bays and rivers, and built dams upriver, such that by the 1970s, writes Sarah Matthews, "any last trace of the wetlands, the lagoon and original course of the river disappeared from the maps completely."[7] Yet Melbourne is persistently amphibious, enlivened by an extensive water infrastructure, periodically shaped by drought, storms, and flooding. Every day the ocean flows into the tidal rivers that cut through the city.

Water is figured as important urban material throughout *A Wrong Turn at the Office of Unmade Lists*. Rawson represents a climate future in which the poor of Melbourne live with the recurrent flooding of the Maribyrnong River in the context of sea-level rise that has altered the geopolitical configuration of Earth and made areas of Australia as well as entire other nations uninhabitable. Occupied by UN peacekeepers who are paid for by and protect the wealthy, Melbourne is riven by deep inequality, from "biofuel refugees" to "carbon credits billionaires."[8] It is a place of discomforting and at times deadly heat, the precarious availability of clean water and other basic resources, and only sporadic provision of public services such as trains and electricity. The novel follows Caddy, barely holding things together two years after her husband and cat died when oil tanks exploded near their house. In the opening pages we are told that a "dirty river" caught fire in a hot wind and that water cannons failed to protect the oil tanks due to a power outage.[9] The references to heat, fire, and breakdown in public services suggest that climate change shaped these deaths, while the narrative of the water cannons highlights tenuous efforts to mobilize water to contain the city's volatile energy infrastructure.

In this story and throughout, *A Wrong Turn at the Office of Unmade Lists* situates water as a "vital matter of city life," to draw on Nikhil Anand.[10] Rawson makes fragmentary but recurring references to water as a crucial element of life in Melbourne. She hints at water's importance even in stories that do not seem to be very much about this element at all, as in her opening reference to how the destruction of Caddy's family is interwoven with the use of water to secure oil infrastructure. This subtle aesthetic attends to the often hidden and indirect ways in which water shapes urban social relations and life possibilities. In *A Wrong Turn at the Office of Unmade Lists* we rarely meet water in a direct and sustained way but rather, to draw on Hugh Raffles, sense it "seeping through at the margins, always there, always in motion, always in mind, and—as if to demonstrate Wittgenstein's observation that 'the aspect of things that are most important for us are hidden because of their simplicity and familiarity'—nearly always invisible."[11]

After the oil tanks explode, Caddy lives with other impoverished and displaced people in the area of Footscray Road, a place that reflects a climate-related resettlement pattern layered onto deeper histories of water-shaped inequality. Footscray is a suburb on the Maribyrnong River in inner West Melbourne with a history of Indigenous displacement and dispossession as well as working-class and immigrant settlement. In the second half of the nineteenth century the river became an industrial corridor and was affected by degradation from slaughterhouses and other facilities: "The factories generated thousands of jobs but gained Footscray the reputation of the smelliest place in Melbourne."[12] The name of Caddy's encampment, Newell Settlement, evokes the urban wetland known as Newell's Paddock, which buffers Footscray when the Maribyrnong River floods. It is projected that a two-meter rise in sea levels will lead to tidal flooding in the riverine areas of Footscray.[13] Indeed, Caddy's home eventually does wash away. "It's not safe here, it floods," she tells a displaced family: "It happens a lot. Floods. The river."[14] Water shapes unequal patterns of settlement, as displaced people find a tenuous refuge in a devalued area prone to flooding.

Water is often imagined in terms of both flow and connection. "Water is truly the transitory element," writes Gaston Bachelard: "Water always flows, always falls, always ends in horizontal death."[15] Yet, as Anand and Ashley Dawson suggest, urban water is also a site where we can trace forms such as constraint, disproportion, and disconnection.[16] Rawson's story of a governmental or corporate failure to blast water through cannons onto a fire is one of elemental flow but also of constraint. "No water got to the cannons," she writes.[17] In the following scene the flow of water is again impeded, as the barkeep Peira tells Jason that he can afford to purchase beer but not water.[18] Jason, whom Peira describes to Caddy as a "kid" who "steals the ice," inhabits a devalued identity as someone who cannot afford to pay for water.[19] He seeks illegal ways of redirecting water to counter his coordinated exclusion from the resource. In a moment of humorous dramatic irony, even as Peira informs Caddy that Jason is an ice thief, we are told that Caddy, too, steals ice, unbeknownst to Peira.[20] Peira's understandings of water are shown to be inadequate, eroding her relatively prestigious social position as a water owner, while Caddy's survival tactics are cast as underhanded. There is a note of comic subterfuge in Caddy's covert efforts to roil the great disproportions bound up in water, in the trickery of gaining an edge on those who hold a large share of this element.

Climate Justice and the Humor of Incongruity

As in the story of the ice thieves, with its emphasis on irony and disproportion, Rawson uses the aesthetic forms of humor to represent urban water, and the broader life in which it is interwoven, in terms of constraint, inequity, and disruption. Much of this humor focuses on Caddy's efforts to survive. The nonperformance of the water cannons is a small but particularly significant detail, reverberating across the entire narrative, for when Caddy's home and family are destroyed, she becomes impoverished, setting in place her troubled effort to get by. This survival takes place especially through her friendship with Ray, who is first introduced as her "friend" and "rock" and who also benefits from "pimping" her.[21] Their relationship is marked by unequal material power, for Caddy relies almost entirely on Ray for resources. In the first scene in which we encounter Ray and Caddy together, she asks him to keep an eye out for her stolen sunglasses, because without them she is exposed to atmospheric dust and harsh sunlight. Ray, an influential figure in the black-market economy, eventually encounters someone wearing the sunglasses and buys them back with the intention of returning them to Caddy. Before he meets Caddy again, however, he comes on an unexpected opportunity to sell them, and his effort to help Caddy takes an unexpected and casual turn toward profit.[22]

Here and throughout *A Wrong Turn at the Office of Unmade Lists,* a subtle humor is built on incongruities, or representations of characters, institutions, and events that seem discrepant, given earlier narratives and context. The humor of incongruity can emerge when one's usual perspectives are thrown into question—when "the world of appearances is turned upside down," to draw from Ralph Ellison.[23] He describes being overwhelmed by laughter on watching Erskine Caldwell's play *Tobacco Road* in New York, having moved from Alabama amid the violence of US racial arrangements in the 1930s. Ellison writes that Caldwell "routed" his composure through a representation of absurdity in the lives of poor southern whites, in turning a common white mode of representing African Americans as "clowns and fools" onto whites.[24] In *A Wrong Turn at the Office of Unmade Lists* incongruity opens up between an earlier narrative—in which Ray intends to help Caddy as her friend—and his quick and casual turn toward profit. The incongruity here takes the form of cosmic irony, a way of thinking that emphasizes forces that exceed and limit human intentions and apprehensions.[25] Both Caddy and Ray fail to fully understand the movement of their lives. For Ray, the very process of attempting to help Caddy sets up conditions for unforeseen economic benefit, while Caddy remains oblivious to the fact that Ray ever recovered and sold her sunglasses. Although these are seemingly trivial details, they disrupt the smooth flow of the narrative, and their humor is laced with more difficult interpretative tensions. Ray's on-the-fly antics are not simply oriented toward economic profit in individualistic terms. They open onto a commentary on the communal and survival, since Ray's accumulation of resources is vital to how he and Caddy get by. Their survival is fractured, embattled, and idiosyncratic, seeming to move in different directions at once, from friendship to exploitation. These ambiguities are further reinforced in that Ray and Caddy do not experience the plight of the sunglasses as funny (or even as incongruous, for that matter), although readers may.

Just as interpretatively disconcerting are the moments in this novel when Ray and Caddy actively mobilize humor toward surviving the challenges of their lives, yet the narrative also dramatizes the complex, even disturbing implications of their humor. Rawson's characters use humor in varied ways, including for keeping some sense of sanity and for momentarily outwitting authorities and overturning power dynamics. Ray uses humor to stabilize his unequal relationship with Caddy. This is evident in a scene in which Caddy hesitates to allow Ray to read a story that she is writing and he mocks her sensitivity: "Yeah, yeah, I know: 'It's silly, you won't like it, blah blah, blah.' Ray was putting on a high-pitched whine which Caddy assumed must be his impression of her."[26] Ray's parody of Caddy directly follows a conversation in which he asks her to have sex with a carbon-credits billionaire he is trying to cultivate a business relationship with. Only after prodding does Ray provide her with cash in return. Here Rawson again illuminates the presence of water in urban social relations, for Ray tries to sweeten the deal by suggesting that the billionaire will pay for a hotel where Caddy can "shower and stuff, clean up on free toiletries," and he also offers her soap because he "want[s] her clean for tonight."[27] The story implies that people's unequal access to clean water has a bearing on their life well beyond the moments when they are directly interacting with that water. Ray wagers that the chance of a shower, a respite from using river water, will make Caddy willing to sleep with the billionaire. His use of humor against Caddy comes after this scene where the resource disparity in their relationship is particularly close to the surface. In mocking her, he expresses affection, closeness, and lightheartedness at a time when these dynamics could be in question. The humor seemingly absorbs something from these tensions, helping them continue in their unequal but interwoven mode of survival. Humor, here, is associated with stability and conservatism, insofar as it maintains some level of acceptance of the inequality in the relationship.[28]

To develop a comic approach to Ray and Caddy—situating them both as unwitting and as active comic creators—is to create an imaginary of future people whose survival is ambiguous and complex. Rawson's humor emphasizes the tremendous challenges that people of this future must survive. The humorous narrative construction highlights rather than minimalizes the difficulty of navigating this future. It is directed to a troubled rather than idealized understanding of communal connections between people in a time of climate change, unequal distribution of freshwater, the disintegration of public life, and severe economic disparity.

The existing theorizations of the humor of incongruity were developed in contexts very different from that of *A Wrong Turn at the Office of Unmade Lists*. Engaging these theorizations alongside this novel is helpful, however, for illuminating how the emphasis in climate justice narratives on incongruities, including disproportion, irony, and absurdity, converges with attention to these forms in humor. Scholars and activists use the concept of climate justice to connect varied struggles that address vast disproportions in people's contributions and exposure to the violence of climate change; some describe this state of affairs as one of "cruel irony."[29] Narratives of climate justice examine race, class, and gender as factors in uneven exposure to climate violence and consider the colonial histories in which fossil fuel industries took form. The relationships between climate and water, in contexts such as sea-level rise, glacial melt, and freshwater insecurity, are a crucial climate-justice issue.

In most narratives of climate justice, the histories and socioeconomic inequalities that shape people's uneven exposure to flooding or to difficulties in acquiring freshwater are unlikely to provoke humor. In *A Wrong Turn at the Office of Unmade Lists*, however, there is the potential to laugh at incongruities in the narrative of a character who is struggling to survive. As I have suggested, the humor of incongruity takes form when we come up against pieces of reality that are impossible to fully reconcile, as when something appears as disproportionate or absurd as against another element of life. Jean Collignon writes that we might imagine a circus clown who attempts to climb a very large wall using a very small chair; the humor of such a scene emerges from its emphasis on disproportion. Yet, Collignon notes, when this clown becomes a prisoner, humorous and tragic potential exist in a disturbing relationship. Collignon describes how Franz Kafka's heroes face unequal power arrangements in works like *The Trial*: "Now nobody will deny that the Kafka hero belongs rather to the category of the prisoner than to that of the clown. And yet, even in his case, there is something comical in the comparison between the wall and the chair."[30] In *A Wrong Turn at the Office of Unmade Lists* disproportion, irony, and absurdity, as sources of humor but also as major elements of climate-justice narratives, come together in a humorous perspective on downtrodden people's efforts to survive climate and water-related upheaval as Rawson emphasizes the controversial and difficult aspects of that survival. Confronting these contexts with humor, Rawson works toward a critique of the socioeconomic inequality interwoven with both urban water and climate change. Such humor emerges from incongruities—albeit in provisional and changing forms—that are not always visible in narratives that situate water in terms of flow and connectivity.

Absurdity and Extinction

Humorous incongruity emerges not only through characters' struggles to survive in *A Wrong Turn at the Office of Unmade Lists* but also through disorienting shifts into absurd narrative elements. Ray, for example, purchases old maps and learns that he can use them to travel through time and space. In addition to making trips to San Francisco, he arrives in a Kafkaesque space called The Gap, the home of bureaucratic offices dedicated to entities like Tupperware Lids, Odd Socks, and Shadows.[31]

The Gap is suggestive of a bureaucratic culture that regulates all manner of seemingly arbitrary things. It evokes the discarded commodities that environmental scholars have sought to reimagine as exerting continued power. Jane Bennett, for example, theorizes the vitality of things that are more often said to be inanimate, implying that materials such as rubbish bear a potentially dangerous agency.[32] In The Gap lost things are carefully cataloged and stored in a space that is not only inaccessible but also unknown to people in Melbourne, perhaps showing the ways in which urban trash is administered out of view in closed places. In such a context, to encounter what one once discarded or otherwise threw off might indeed feel surreal.

Less predictable things that have been lost also accumulate in The Gap. One office collates shadows; another collects disrupted dreams. Yet another, dedicated to Unmade Lists, is expanding because there are "so many lists to not yet make."[33] A worker tells Ray that one of the unmade lists is called "Smells Now Extinct" and speculates about the smells that might be on such a list:

> One. The smell of a trilobite which has been feeding underwater and is now half exposed by a receding tide.
>
> Two. The smell of a trilobite which has been feeding underwater and has become wedged between two rocks. In its struggle to escape, part of its exoskeleton has been crushed.
>
> Three. The smell of three brine shrimp which have smelled the crushed exoskeleton and have come to investigate. One is nibbling on the trilobite's exposed flesh.[34]

Trilobites are marine arachnomorph arthropods that survived for about 270 million years and are now known for their extensive fossil record; some have been found in metropolitan Melbourne. Trilobites became extinct about 252 million years ago during the Permian-Triassic extinction event, in which it is thought more than 90 percent of marine species died out through some combination of volcanic activity, ocean acidification, marine anoxia (lack of oxygen), hypercapnia (high carbon dioxide), increased sediment runoff, and severe global warming.[35] In *A Wrong Turn at the Office of Unmade Lists* references to these ancient beings are juxtaposed with a narrative primarily about urban life, evoking more recent urban-oceanic extinctions, such as that of the "functionally extinct" shellfish reefs of Port Phillip Bay, where Melbourne is situated.[36] The bay's blue mussel and native flat oyster reefs were destroyed through dredge fishing for food; the oysters also became a source of the lime used in mortar for the colonial building industry.[37]

In approaching the histories of extinction in terms of an unmade list of vanished smells, *A Wrong Turn at the Office of Unmade Lists* plunges us into a world where what is valued and curated is highly unusual, even absurd. It perhaps calls to mind an early European Australian literary history of humor, described by Fran De Groen and Peter Kirkpatrick in terms of "larger-than-life fantastic elements and a certain imaginative exuberance," such as a story in which a sheep's wool grows so high that a ladder is required to shear it.[38] The fantastic workings of humor can still be seen in European Australian cultural production, including in that about climate change. George Miller's film *Mad Max: Fury Road* (2015), for example, evokes climatic and radioactive catastrophes through such offbeat details as a flame-throwing heavy-metal guitarist who accompanies leader Immortan Joe and the war boys across the desert.[39] Sally Abbott's novel *Closing Down* (2017), in which Australians and people globally are displaced to "inclusion zones" in a future time characterized by a changing climate and many other strange phenomena, is primarily somber but is tinged with moments that are absurdly humorous.[40] In one scene a gardener at a Nairobi hotel hoses plastic plants and sprays them with various scents, explaining that the hotel no longer has enough water to sustain plants but that a plastic garden can be watered and the water subsequently caught and recycled: "We can water the plastic forever . . . which is rather nice if you stop to think about it. . . . A real garden can be a melancholy thing."[41] Not only do privileged people consume the performance of watering a garden that does not need water, but they also seek to maintain some semblance of their former world—a garden that requires water—through the fossil fuels, commonly used in plastics, that are a factor in that world's undermining.

While Ray is described as an Indigenous character in *A Wrong Turn at the Office of Unmade Lists*, *Closing Down* and *Mad Max: Fury Road* do not engage with Indigenous peoples in their imagined futures. In contrast, the Waanyi nation writer Alexis Wright's novel *The Swan Book* (2013) uses satiric humor to link the persistent undermining of Indigenous life and sovereignty to a world devastated by a changing climate and wars.[42] The novel is suggestive of the Eualeyai/Kamillaroi scholar and writer Larissa Behrendt's argument that many Indigenous Australian writers, comedians, and filmmakers use laughter as a healing balm and an educational tool against the devastation of racism.[43] Behrendt speaks with stand-up comedian Kevin Kropinyeri, who suggests how humor plays out differently in relation to different audiences: "I love performing for an all black audience. I don't have to explain things."[44] To interpret a comic mode is to do just that, in a way relative to the experiences and cultural assumptions we bring. Indeed, Jane Gleeson-White notes the "apparent blend of real and fantastic" in *The Swan Book* but argues against Western critical binaries, including the concept of magical realism, which "fails to account for the complex *reality* of the world that Wright endeavours to bring to fiction."[45] A narrative read within certain Western critical contexts as incongruous in its seemingly fantastic elements may not be read in this way in a context of different critical assumptions about the scope of reality and its usual modes of narration.

A Wrong Turn at the Office of Unmade Lists is one of a number of Australian works of climate fiction that see possibilities of humor in relation to climate change, though that humor emerges within distinct traditions, but the novel more directly calls up transnational affinities that are carried in its absurd humor, referencing Kafka in association with The Gap.[46] Michael Y. Bennett describes the tragicomic creativity of the literature of the absurd, a literature influenced by Kafka and especially vibrant from the 1950s through the 1970s. Bennett argues that this literature often places characters in a uniquely ridiculous situation, quoting from *Oxford English Dictionary*'s definition of *ridiculous* as "exciting ridicule and derisive laughter; absurd, preposterous, comical, laughable."[47] Such texts call up "a tragicomic response: one must both laugh and cry, sometimes simultaneously, sometimes alternately, and sometimes one is unsure which response is adequate and/or appropriate."[48] The most outlandish works of this literature provide a close-to-the-bone critical engagement with how reality is understood.

Indeed, what at one level in Rawson's novel may seem laughable—even ridiculous—also offers a very tragic perspective on the ocean. The losses attending climate change include not simply individuals or species but also myriad relationships and experiences, such as smells. Among the many ridiculous features of Rawson's novel is that a brine shrimp is eating a trilobite even though brine shrimp had not evolved at the time the trilobites were alive. The list of vanished smells includes not only those that may have become extinct but also those that never existed, because they were foreclosed through extinction. It describes smells that *might* have been if the trilobites had survived to the point at which the brine shrimp took form. Extinction reverberates in ways that cannot fully be apprehended, changing not only present life but also its futures.

Layers of interpretative potential are interlaced here: the unmade list is absurdly laughable but also speaks to the abyss of loss that is the interconnection of the oceans and climate change. To draw on Glenda R. Carpio's work on tragicomedy, the potential here is for uneasy, obstructed laughter. Reading the work of Charles W. Chesnutt, Carpio describes a form of humor in which "one is caught between a desire to laugh and the suspicion that, in doing so, one could be cruelly laughing at a tragedy that is about to unfold. One is caught, that is, between wanting and not wanting to laugh."[49] She argues that these tensions suggest suffering and violence that exceeds representation. In *A Wrong Turn at the Office of Unmade Lists* the humorous, deeply sad, and critical aspects of the material are impossible to fully reconcile, turning us toward the violence of climate change that, at some level, is impossible to represent. This unaccountable, disturbing aesthetic speaks of such violence but without the imaginative closure that could elide the diverse experiences that are of concern in climate-justice narratives.

Comic Humor and Petroleum Energy Systems

In imbuing climate change with a sense of absurdity, Rawson not only sketches out ways in which it troubles representation but also suggests that it is an unnecessary catastrophe. She makes this point in a confrontational scene that begins and ends abruptly, in which Xotchel, an impoverished eleven-year-old who survives by selling manicures on the streets, is killed. This short scene opens as Xotchel sits and watches a group of "fish-gilled men" disembark from a boat on the river.[50] As one of the men exits the boat, he moves toward her, sinks his teeth into her ankle, and takes off her foot. In the last reference to Xotchel in the novel, "she stared him hard in the eyes as he gnawed at her exposed bone and then she lay back on the burning cement."[51] Merging elements of horror and absurdity, the scene again calls to mind the novel's tragicomic tensions and representations of violence that nevertheless also remain excessive to representation. As I have argued in this essay, absurd humor might be used to represent the unequal violence of climate change even while illuminating the inadequacy of such representation.

At one level, Xotchel's death may reflect a theme explored throughout *A Wrong Turn at the Office of Unmade Lists*: the seething horrors of imagination and memory amid the violence of climate change. In the world of this novel, what has been imagined or remembered persists, even when it is discarded or forgotten, and can seep back into the experiences of the living, taking on a disturbingly material and lively form. For example, Caddy's dead husband, Harry, and cat, Skerrick, both killed when the oil tanks exploded, walk back into Caddy's life many years later. Caddy, who has been haunted by memories of them, tells Harry that she believes that she called him back: "I imagined you here."[52] Harry, who returns in an altered and precarious form, tragically melts away into a shadow and once again is lost to Caddy. Although Rawson does not directly explain the appearance of the fish-men in these terms, as another instance of an imaginary that gains a frightening materiality, it is possible that these nightmarish beings are dreamed up by Xotchel or someone else and eventually do gain substance. Rather than only emphasizing future people's material struggles to survive in the context of climate change, Rawson speculates on the horror of their memories, dreams, and imaginaries, haunted as they are by relationships with the dead and by visions of further violence to come.

Although the fish-men may be a speculation on the violent liveliness of imagination and memory amid the traumas of climate change, these beings also directly link petroleum energy systems and the ocean to the future imagined in *A Wrong Turn at the Office of Unmade Lists*. As they spill off the boat, they "shimmered with electricity and their black-green skin oozed sparkling oil."[53] They embody a fishy materiality, with their gills, blackish-green color, oiliness, and the capacity to produce electrical fields. Such language simultaneously associates the fish-men with capitalist energy systems, particularly with oil. The strange beings symbolize the uneven interconnection of humans, the ocean, and petroleum energy, as well as the violence bound up in this interconnection. Insofar as the fish-men are slathered in oil and can make electricity, Rawson represents the control and use of petroleum energy in terms of gendered structural inequality. In Australia, people's relationships with the responsibility and violence of the fossil fuel industries are highly unequal.[54] Xotchel's foot is ripped off her body and eaten, suggesting the commodification of girl's and women's bodies as they are displaced and impoverished in the context of climate change, a theme that Rawson also explores through Caddy's sex work. Rawson critiques fossil fuel capitalism in emphasizing the absurdity—the extremity, ridiculousness, and senselessness—of the violence it is shaping.

Although Rawson critically examines the violence of petroleum capitalism through the narrative of fish-gilled men, we might also remember the persistent element of ridiculousness in this narrative. If, to draw on Bennett, "it is . . . ridiculous to say that there is not an element of the *ridiculous* in absurd literature," Rawson's narrative of electric, oily fish–human beings that arrive in Melbourne by boat is decidedly ridiculous.[55] Its humor might be understood as comic, in the sense that it targets not only climate change but turns on its own, in certain ways ridiculous narrative. One of the few scholars to examine the potentials of humor for ecocritical and environmental thought, Nicole Seymour highlights comic humorous approaches that turn with uncertainty on their own engagement and that allow for a relationship of curiosity, rather than of certainty, with objects and audiences.[56] Humorously insisting on the uncertainties that infuse its narrative, *A Wrong Turn at the Office of Unmade Lists* clears space for varied interpretations and to recognize silences, as in this scene where Xotchel stares without speaking as she is killed. This comic approach reminds us that the work of finding ways to survive climate change lies in shared struggles that are not contained by a narrative.

In these contexts of water and climate-related precariousness, laughter can seem to be almost as strange and turbulent as the world it engages. Yet hydro-critical reading practices should perhaps be attuned to humor among the modes capable of engaging water in artful and critical ways. Rawson's techniques of incongruity situate urban water environments in forms that include not only flow but also intricate patterns of constraint, disproportion, and disruption. Her emphasis on ridiculousness attends to the absurd but real processes of climate change and to the challenges of their imagining and remembrance, with earnest but no less comic creativity.

TERESA SHEWRY is associate professor of English at the University of California, Santa Barbara. She specializes in the environmental humanities, Pacific literatures, utopian literatures, and theories of humor. She is author of *Hope at Sea: Possible Ecologies in Oceanic Literature* (2015).

Notes

1 The NOAA projects global mean sea-level rise of 2.5 meters by the end of this century in a worst-case scenario involving business-as-usual emissions trajectories and taking into account potentially intensified ice melt. See Sweet et al., "Global and Regional Sea Level Rise Scenarios," 21.

2 The video is widely available online. See Keaney, "Peter Dutton Overheard Joking."

3 Hasham, "Immigration Minister."

4 Lucariello, "Situational Irony," 129.

5 Matthews, "Erasure of Melbourne's Wetlands."

6 Robertson, "Journey from Swampy Badlands."

7 Matthews, "Erasure of Melbourne's Wetlands."

8 Rawson, *Wrong Turn*, 118, 73.

9 Rawson, *Wrong Turn*, 9.

10 Anand, *Hydraulic City*, 1. Anand, who examines how marginal groups in Mumbai get by through accessing the city's hydraulic infrastructure, is among several social scientists and humanists who focus on water as a crucial and contested urban material. John F. López explores the histories of water in Mexico City and reconstructs "how water shaped the settlement's spatial organization and urban fabric" ("In the Midst of Floodwaters," 35). Ashley Dawson explores the vulnerability of cities in relation to sea-level rise, arguing that the science of climate change too often "ignores the specific places where most of us live—cities" (*Extreme Cities*, 7).

11 Raffles, *In Amazonia*, 181.

12 Barnard, "Maribyrnong River Heritage Trail," 3.

13 A speculative mapping of such inundation is available on the Coastal Risk Australia 2100 website, coastalrisk.com.au (accessed July 2, 2018).

14 Rawson, *Wrong Turn*, 242.

15 Bachelard, *Water and Dreams*, 6. See also Elizabeth DeLoughrey's references to varying ways in which ideas of fluidity have been associated with the ocean in contexts of territorialization, resource extraction, and decolonization ("Submarine Futures").

16 Dawson suggests such forms in describing a "wall of water" that flowed through Manhattan during Hurricane Sandy, revealing the "disconnected and disjointed character" of the city as subways came to a halt, power was cut, and some neighborhoods were left without drinking water (*Extreme Cities*, 2, 3). Anand draws on James Ferguson to differentiate disconnection from water, an actively produced deprivation of previously held access to social and political systems, from the state of being unconnected to water, in which subjects never had access to those systems (*Hydraulic City*).

17 Rawson, *Wrong Turn*, 9.

18 Rawson, *Wrong Turn*, 16.

19 Rawson, *Wrong Turn*, 13.

20 Rawson, *Wrong Turn*, 13.

21 Rawson, *Wrong Turn*, 24, 49, 51.

22 Rawson, *Wrong Turn*, 60.

23 Ellison, "Extravagance of Laughter," 146.

24 Ellison, "Extravagance of Laughter," 194, 185.

25 See Colebrook, *Irony*, 14.

26 Rawson, *Wrong Turn*, 53.

27 Rawson, *Wrong Turn*, 52, 71.

28 Angelique Haugerud describes the contentious debates around the extent to which humor performs a conservative social function. From the perspective of those who view humor as conservative, "far from being threatening, political humor actually contributes to social stability, redirecting emotions people might otherwise channel to revolution." She argues that rather than try to decisively link humor to certain outcomes, we must consider its implications in specific contexts (*No Billionaire Left Behind*, 30).

29 On the "cruel irony" of climate injustice, see Cole, "Climate Change's First Victims." Brian Tokar writes that "climate justice embodies the fundamental understanding that those who contribute the least to the excess of carbon dioxide and other greenhouse gases in the earth's atmosphere consistently and disproportionately experience the most severe and disruptive consequences of global warming, and are often the least prepared to cope with its consequences" (*Toward Climate Justice*, 25).

30 Collignon, "Kafka's Humor," 56.

31 Rawson tells us that the maps were created on July 16, 1945, the day that the first nuclear weapon, Trinity, was detonated. Further references link The Gap to the United States. For example, a worker in The Gap speaks with an American accent and offers Ray a card that lists a gov/dssr URL, evoking the Department of State Standardized Regulations, a US Department of State agency that manages allowances and benefits for US Government civilians overseas. These references suggest often-hidden transnational connections that shape the world of Caddy and Ray, including ways in which US militarization and foreign policy have secured processes of extractivism that underlie anthropogenic climate change (*Wrong Turn*, 207–8, 84).

32 Bennett, *Vibrant Matter*, viii.

33 Rawson, *Wrong Turn*, 237.

34 Rawson, *Wrong Turn*, 96.

35 See Song et al., "Two Pulses of Extinction."

36 Ford and Hamer, "Forgotten Shellfish Reefs," 93.

37 Ford and Hamer, "Forgotten Shellfish Reefs," 90.

38 De Groen and Kirkpatrick, "Introduction," xxi.

39 *Mad Max: Fury Road*, directed by George Miller (Warner Brothers, 2015), DVD.

40 Abbott, *Closing Down*, 17.

41 Abbott, *Closing Down*, 121–22.

42 Wright, *Swan Book*.

43 Behrendt, "Aboriginal Humour."

44 Behrendt, "Aboriginal Humour."

45 Gleeson-White, "Going Viral."

46 Rawson, *Wrong Turn*, 210.

47 Bennett, *Cambridge Introduction*, 10.

48 Bennett, *Cambridge Introduction*, 19.

49 Carpio, *Laughing Fit to Kill*, 36.

50 Rawson, *Wrong Turn*, 225.
51 Rawson, *Wrong Turn*, 225.
52 Rawson, *Wrong Turn*, 309.
53 Rawson, *Wrong Turn*, 225.
54 For example, the Australian government's Workplace Gender Equality Agency (WGEA) Data Explorer estimates that in 2017 the Australian electricity supply workforce was 26.8 percent female and the gas supply workforce was 26.2 percent female. See Australian Government, "WGEA Data Explorer."
55 Bennett, *Cambridge Introduction*, 10.
56 Seymour, "Toward an Irreverent Ecocriticism," 62.

Works Cited

Abbott, Sally. *Closing Down*. Sydney: Hatchette, 2017.

Anand, Nikhil. *Hydraulic City: Water and the Infrastructures of Citizenship in Mumbai*. Durham, NC: Duke University Press, 2017.

Australian Government. "WGEA Data Explorer." Australian Government Workplace Gender Equality Agency, 2018. data.wgea.gov.au/overview (accessed November 25, 2018).

Bachelard, Gaston. *Water and Dreams: An Essay on the Imagination of Matter*, translated by Edith R. Farrell. Dallas, TX: Pegasus Foundation, 1983.

Barnard, Jill. *Maribyrnong River Heritage Trail*. Melbourne: Maribyrnong City Council, 2008.

Behrendt, Larissa. "Aboriginal Humour: 'The Flipside of Tragedy Is Comedy.'" *Guardian*, June 19, 2013. www.theguardian.com/commentisfree/2013/jun/19/aboriginal-comedy-humour.

Bennett, Jane. *Vibrant Matter: A Political Ecology of Things*. Durham, NC: Duke University Press, 2010.

Bennett, Michael Y. *The Cambridge Introduction to Theatre and Literature of the Absurd*. Cambridge: Cambridge University Press, 2015.

Carpio, Glenda R. *Laughing Fit to Kill: Black Humor in the Fictions of Slavery*. Oxford: Oxford University Press, 2008.

Cole, Lily. "Climate Change's First Victims Are Always Those Least to Blame." *Guardian*, December 3, 2014. www.theguardian.com/commentisfree/2014/dec/03/climate-change-yanwana-amazon-floods-lily-cole.

Colebrook, Claire. *Irony: The New Critical Idiom*. London: Routledge, 2004.

Collignon, Jean. "Kafka's Humor." *Yale French Studies*, no. 16 (1955): 53–62.

Dawson, Ashley. *Extreme Cities: The Peril and Promise of Urban Life in the Age of Climate Change*. London: Verso, 2017.

De Groen, Fran, and Peter Kirkpatrick. "Introduction: A Saucer of Vinegar." In *Serious Frolic: Essays on Australian Humour*, edited by Fran De Groen and Peter Kirkpatrick, xv–xxvii. Queensland: University of Queensland Press, 2009.

DeLoughrey, Elizabeth. "Submarine Futures of the Anthropocene." *Comparative Literature* 69, no. 1 (2017): 32–44.

Ellison, Ralph. "An Extravagance of Laughter." In *Going to the Territory*, 145–97. New York: Random House, 1986.

Ford, John R., and Paul Hamer. "The Forgotten Shellfish Reefs of Coastal Victoria: Documenting the Loss of a Marine Ecosystem over 200 Years since European Settlement." *Proceedings of the Royal Society of Victoria* 128 (2016): 87–105.

Gleeson-White, Jane. "Going Viral: *The Swan Book* by Alexis Wright." *Sydney Review of Books*, August 23, 2013. sydneyreviewofbooks.com/going-viral.

Hasham, Nicole. "Immigration Minister Peter Dutton Caught Joking about the Effect of Climate Change on Pacific Islands." *Sydney Morning Herald*, September 11, 2015. www.smh.com.au/federal-politics/political-news/immigration-minister-peter-dutton-caught-joking-about-the-effect-of-climate-change-on-pacific-islands-20150911-gjkfoz.html.

Haugerud, Angelique. *No Billionaire Left Behind: Satirical Activism in America*. Stanford, CA: Stanford University Press, 2013.

Keaney, Francis. "Peter Dutton Overheard Joking about Rising Sea Levels in Pacific Islands Nations." ABC News, September 11, 2015. www.abc.net.au/news/2015-09-11/dutton-overheard-joking-about-sea-levels-in-pacific-islands/6768324.

López, John F. "In the Midst of Floodwaters: Mapping Viceregal Mexico City's Urban Transformation, 1524–c. 1690." *Rutgers Art Review* 28 (2012): 35–52.

Lucariello, Joan. "Situational Irony: A Concept of Events Gone Awry." *Journal of Experimental Psychology: General* 123, no. 2 (1994): 129–45.

Matthews, Sarah. "The Erasure of Melbourne's Wetlands." *State Library Victoria: What's Your Story* (blog), August 29, 2016. blogs.slv.vic.gov.au/such-was-life/the-erasure-of-melbournes-wetlands.

Raffles, Hugh. *In Amazonia: A Natural History*. Princeton, NJ: Princeton University Press, 2002.

Rawson, Jane. *A Wrong Turn at the Office of Unmade Lists*. Melbourne: Transit Lounge, 2013.

Robertson, Jennifer. "The Journey from Swampy Badlands to Wetlands." *Scientific Scribbles* (blog), August 20, 2016. blogs.unimelb.edu.au/sciencecommunication/2016/08/20/the-journey-from-swampy-badlands-to-wetlands.

Seymour, Nicole. "Toward an Irreverent Ecocriticism." *Journal of Ecocriticism* 4, no. 2 (2012): 56–71.

Song, Haijun, Paul B. Wignall, Jinnan Tong, and Hongfu Yin. "Two Pulses of Extinction during the Permian-Triassic Crisis." *Nature Geoscience* 6, no. 1 (2013): 52–56.

Sweet, William V., Robert E. Kopp, Christopher P. Weaver, Jayantha Obeysekera, Radley M. Horton, E. Robert Thieler, and Chris Zervas. "Global and Regional Sea Level Rise Scenarios for the United States." NOAA: National Oceanic and Atmospheric Administration Technical Report NOS CO-OPS 083. Silver Spring, MD, 2017. tidesandcurrents.noaa.gov/publications/techrpt83_Global_and_Regional_SLR_Scenarios_for_the_US_final.pdf.

Tokar, Brian. *Toward Climate Justice: Perspectives on the Climate Crisis and Social Change.* Porsgrunn: New Compass, 2014.

Wright, Alexis. *The Swan Book*. New York: Atria, 2013.

Hydro-eroticism

JEREMY CHOW AND BRANDI BUSHMAN

Abstract In tracing the ways that eroticism and violence are mapped onto bodies of and relationships with water, this essay offers "hydro-eroticism" to consider an ecofeminist and queer ecological reading of water. *Hydro-eroticism* signifies two interventions: first, how aqueous locations become sites of queer community and punishment while registering associations with the fluid female body, and second, how human and nonhuman intimacy is enabled by aqueous proximity. The essay focuses on Hulu's *The Handmaid's Tale*, based loosely on Margaret Atwood's novel, and *The Shape of Water*. Whereas in *The Handmaid's Tale* the river is configured doubly as a site of bodily violation wherein violence against queer characters is palpable but also as a space that informs nostalgic reunion for queer female community, in *The Shape of Water* hydro-eroticism speaks to the fraught and layered entanglements of human female and nonhuman male creature in concert with queer kinship.

Keywords water, ecofeminism, queer ecology, sexuality, violence

If, as Stacy Alaimo poignantly contends, "the anthropocene is no time to set things straight," then *now* is the time to weigh our exposure to environmental interactors that might unsettle normativity altogether.[1] *Exposed*, Alaimo's book whence this playful opening line hails, examines the forms of environmental pleasures that might accompany human and nonhuman bodies, the blending, blurring, and bleeding of what she earlier theorized as "trans-corporeality."[2] Deeply attuned to these pleasurable shifts and metamorphoses, we offer "hydro-eroticism" as a hermeneutic that mediates the intersecting vertices wherein eroticism and violence meet in bodies of and relationships with water. Employing ecofeminist and queer ecological readings of water and the female body, we attend to water's feminist roots and queer potentials. For example, in her indispensable feminist manifesto, *The Second Sex*, Simone de Beauvoir conjures an early mythic history that presupposes a gendered binary for bodies of water. For Beauvoir and the Babylonians about whom she makes this analysis, the ocean is male and the sea is female, and thus "a single element often has an incarnation that is at once male and female . . . the double incarnation of cosmic chaos."[3] The doubled gendering—the perhaps trans affinity—of elemental water invokes Beauvoir's articulation of a cosmic chaos:

57:1, April 2019 DOI 10.1215/00138282-7309710

worlds vexed, universes panicked. In another established mythic cosmos, illustrations of Charybdis and Scylla invoke toothy, insatiable maws that carry misogynist associations with the *vagina dentata,* Madea, and other uncontrollable women whose female prowess and appetites are obstacles to overcome for patriarchal pioneers. The vilified female body and its connection with water are unmistakable, and as these myths imply, there is no escaping the violence that accompanies the sensuous sea. Literary canons and oral mythic histories remind us to heed this warning.[4]

In conceiving of "hydro-eroticism," we emphasize erotic connectivities that bodies of water foster, while also remaining mindful of how fluid affiliations are policed, inhibited, or persecuted. Aqueous entanglement offers both a queer erotic potential and, often in the same moment, the foreclosure of that potentiality. Like aqueous bodies themselves, we highlight the fluid, the multiple, and the mutable, which speaks to our investment in a capacious ecofeminism and queer ecology, fields that are not identical but work in tandem as harbingers of aqueous thinking. "Hydro-eroticism" thus undertakes three central interventions: first, how aqueous locations become sites of queer community and punishment; second, how these sites register associations with the fluid female body; and third, how human and nonhuman intimacy—the erotic nuance of sex thought impossible and prohibited—is enabled by aqueous proximity and immersion. We examine these aquatic potentials by pairing Hulu's smash hit *The Handmaid's Tale,* based loosely on Margaret Atwood's 1986 Governor General's Award–winning novel, and the Oscar-winning film *The Shape of Water.* In these two filmic adaptations, we do not see identical forms of hydro-eroticism operating. Hydro-eroticism is not about limitation; instead it dwells in the multifold, the myriad, the impossibly entangled intimacy with and within water. In *The Handmaid's Tale* the river is configured doubly, not only as a site of bodily violation wherein violence against queer characters is palpable but also as a space that informs nostalgic reunion for queer female community. The fluid pleasure of queer eroticism is sensually grown with water's participation, while queer female pleasure is beset by heteromasculine anxiety, which induces a violent brutalization of the body. Separately, in *The Shape of Water* hydro-eroticism bespeaks the fraught and intensely layered entanglements of human female and nonhuman male creature in concert with animality, sex, and queer kinship. In both examples we locate the erotic's proximity to violence, which offers a queer eroticism that works with, through, and against modes of violence. Our goal is not to prescribe a particular ontology or teleology,[5] which might suggest a limited and inherent quality to water, potentially undermining our investments in gender and sexuality studies. Rather, we intend to envision how two contemporary media sensations further invoke both a gendered and queered affinity with water—another way to conceive of the plural articulations of "hydro-power" elaborated on in this collection—which can grow the intersections of the environmental humanities, especially with the Oceanic Turn or the Blue Humanities,[6] with gender and sexuality discourses.

To study eroticisms near and with water, it is necessary to cast off presumptions of water as something nonagentic, inert, and therefore exploitable. Here we explore a perspective that considers water lively and active, specifically in mediating intimacies between humans, and between humans and nonhuman species. Seeing water as an active *participant* in mediating intimacies and creating pleasure—a third party, perhaps—heightens the stakes of queer ecocriticism. Alongside new materialist scholarship, we look to water's participation in the sexual liaison, which coincides with what Karen Barad refers to as "intra-activity" and addresses the porous nature of human and nonhuman bodies, an understanding echoed by Astrida Neimanis's vivid articulation of aqueous potential as that which "rupture[s]" and "renegotiate[s]" "seething" bodies.[7] Hydro-eroticism homes in on these porous boundaries, varied participations, and sites of sexuality by accepting fluid bodies and identities that subsequently reveal the possibilities for intimate pleasure with and through water. The interdependency of the human body, its own fluidity, and external waters can enhance pleasure, yet the violence ever surrounding queer intimacies can render these pleasurable encounters precarious. To counter this precarity, we wager that considering water and aquatic, nonhuman beings as entities with whom we can commune in erotic intimacies unleashes new pleasurable potential. As Donna Haraway states in *When Species Meet*, "To be one is always to *become with* many."[8] Haraway's mode of becoming augurs the fluidity of queer eroticism: a revelation that welcomes the vibrancy of plural intimacies alongside ecological agents such as water.

By dwelling in a fluid analysis, we acknowledge that fluidity can also signal a source of heterosexual, patriarchal anxiety, feelings that foreground the oppression of queer bodies and intimacies as "unnatural." Such anxieties and oppressions are integral to *The Handmaid's Tale* and resonant with *The Shape of Water*. The appropriation and use of "nature" as a means to ostracize the "unnatural" queer body is noted by ecofeminist and queer ecology scholars like Greta Gaard, Catriona Mortimer-Sandilands, Bruce Erikson, and Nicole Seymour.[9] While "nature" is often wielded against queer sexuality, nature and women are linked as inherently close with shared identities, essentialisms that are similarly used as means of dehumanizing women and ravaging nature. Gaard notes that "nature and all that is associated with nature, including women, the body, emotions, and reproduction" are detrimentally linked and thereby oppressed.[10] These equations have forged an ideological conflation between femininity (especially the maternal body and female genitalia) and water in contradictory modes; while idealized and fetishized female purity and motherhood conjure associations with clean waters, female sexuality and genitalia have often been linked to abject, swampy waters.[11] In considering the connections among women, nature, and water, we honor an ecofeminist perspective of eroticism. In "Toward a Queer Ecofeminism" Gaard aligns ecofeminism and queer ecology to pinpoint eroticism as a root of queer oppression located in "the erotophobia of Western culture, a fear of the erotic so strong that only one form of sexuality is overtly allowed; only in one position; and only in the context of certain legal, religious, and social sanctions."[12] It is this emphasis on eroticism that hydro-eroticism follows.

Both *The Handmaid's Tale* and *The Shape of Water* enfold the white female body in violently erotic potentialities. Despite our archive, hydro-eroticism is not singularly a vehicle for reading white female eroticism; critical race and postcolonial readings of this same hermeneutic are ripe for exploration. Our undertaking of hydro-eroticism is carefully attuned to the fluidity of intimacies that fluctuate among female and queer or nonhuman bodies, and, in so doing, we celebrate the fluidity of queer and plural intimacies grounded in varying identities, relations, and embodied presentations. Alongside new materialist discussions of animacy, our work is similarly attentive to Indigenous perspectives of animism. Mel Chen describes animacy as "a quality of agency, awareness, mobility, and liveliness,"[13] while Eduardo Kohn heightens these stakes by stating that "selves, not things, qualify as agents."[14] In *How Forests Think* Kohn notes that the animism of the Amazonian Runa is more than a foil to Western perspectives on the environment; however, director Guillermo Del Toro pointedly contrasts the Amazonian river creature in *The Shape of Water* with the white, Christian, heterosexual government employee tasked with torturing the creature to discover its "secrets." We discuss this cultural blending and rupture below. Rather than appropriating Indigenous modes of thinking, we examine the tensions that arise from Judeo-Christian perspectives that condemn queer intimacies and infantilize Indigenous perspectives to show that hydro-eroticism can push against both Western culture and heteromasculinity.

Because hydro-eroticism calls for thinking through relations that are often oppressed or marginalized and therefore invisible to dominant, mainstream perception, our study arises out of a need to speculate and imagine other intimacies, possibilities, and beings. Atwood's dystopia and Del Toro's fantastic fairy tale allow us to exemplify Alaimo's call for increased speculation in ecocritical thinking, a need to "unmoor" primarily "by practicing informed, intentioned speculation about other creature's perspectives, modes of beings, and lifeworlds."[15] Informed speculation is depth oriented, imaginative, and immune to boundaries and regulations, perspectives necessary to unveil the queer potential in erotic and violent aquatic communion. To unmoor from dominant methods of thinking about and interacting with the environment—to turn toward a blue ecological perspective—there must be, in Alaimo's words, "an intentional letting go in order to loosen the confines of the human."[16] Dwelling in the aqueous, hydro-eroticism offers a reorientation of how we view and commune with bodies of water, a deconstruction of confines that limit intimacies, and a (necessary) queering of the Anthropocene that welcomes the pleasures of aquatic intra-action.

Temporal Living, Violent Foreclosures, and Homoerotic Fluidity

Hydro-eroticism suggests that water's polyvalent identity, much like its ebbs and flows, is tempestuous, sensuous, and violent. Such aqueous power informs female erotic relations—and parallels femininity and queer sexuality—in Hulu's 2017 adaptation of *The Handmaid's Tale*, a rendition that amplifies Atwood's world of horrors. The adaptation works diligently to envision the representation of queer individuals in the dystopic Gilead, exposing the oppressive hegemony of a fundamentalist, heterosexual-enforcing society against queer individuals. Atwood's speculative future predicts the enslavement of fertile women: ecological degradation and industrialized toxicity irreparably damage the "healthy" human body, making successful births a rarity. Fertile women are enslaved as "handmaids"; as such, they are raped monthly by high-ranking Commanders of Gilead in the hopes of repopulating society. Fertile women who dissent are punished by hanging or removal either to the colonies—described in Atwood's novel as a site of "toxic dumps and radiation spills" with contaminated water and polluted air—or to "Jezebel's," a former hotel turned brothel for the Commanders' sexual appetites not sated by their wives or handmaids.[17] In the series, aqueous spaces such as the river, the bathroom, and the polluted water in the colonies all situate queer intimacies amid violence. Precarious, short-lived, or volatile, the intimacies forged in aqueous sites present different bodies of water as brief fulfillments of queer sexual desires that are ultimately policed and punished.

Gilead uses aqueous sites to punish queer individuals, as the river is appropriated as both a site of transgression and a showcase for punished queer bodies.[18] As respite from the despotic, patriarchal city space, Offred (Elizabeth Moss) and Ofglen (a queer handmaid played by Alexis Bledel) walk around the city and trace the riverbank with their steps. During these walks Offred and Ofglen ultimately recuperate the punitive river space as one wherein female intimacy can flourish. Once remade as a refuge from a cityscape that ideologically and physically enforces heteronormativity, the series employs the river as a space for community that allows for a fleeting indulgence in nostalgia and hope. Offred and Ofglen forge an intimate relationship with one another and with the river itself, reclaiming the water space by subverting patriarchal and heteronormative ontologies that reject queer people from "'nature' or natural spaces."[19] The meshed intimacy follows Neimanis's concept of an "inextricable entanglement" between humans and the environment.[20] "Human ideas, meanings, and values," Neimanis states, "shape and are shaped by, in some important way, the 'environment out there,'" an assertion that we extend to suggest that female and queer intimacies not only are shaped by the environment but also use and *depend on* water specifically for construction and fulfillment.

As Offred and Ofglen commence their walk alongside the river, the visible reminder of punishment enacted by the state is unmistakable: over a stone wall, surrounded by swarming flies and banks rife with weeds, lawbreaking individuals are hanged for their transgressions. Offred notes that one of these bodies is a "gay man," marked by the pink triangle on the burlap sack covering his head, a historical allusion to the markedness of queer bodies during the Shoah.[21] The dilapidating, cracked white wall evokes brutality and violence, while its discolored hues of gray and black physically link the way in which Gilead ideologically associates queer sexualities with deviance, impurity, and sin.[22] The gay man's body is not only punished by death but, perhaps more perniciously, is also dispensed with as degenerate waste, as it rots, covered in dirt, swarmed by flies. The queer body becomes detritus. Gilead associates the queer body with dirt and sin, physically ensuring that the dead corpse is tainted by the abject environmental material to which it is remanded. While dirt symbolizes abjection, the river itself stands in for abjection, as these waters have a cultural history that invokes transgression.[23] Jeffrey Jerome Cohen specifically notes the impure nature of these waters in his description of the river: "a place of interstices, mixing, hybridity, autonomy, cogency. The closer to the sea it flows the more impure it becomes, culminating in an estuary that combines saltwater with fresh and everything with mud."[24] Extending Cohen, we posit the river as an ecological body that invites queer ontologies in that it actively resists and subverts predictable, normative, static behavior and definition. The man's proximity to the riverbank, the intersection of land and water, suggests an intimacy between queer sexuality and aqueous space. Both are rendered abject in *The Handmaid's Tale*. It is in proximity to the river that the man is punished and hanged like an effigy, and thus hydro-eroticism enables us to visualize how aqueous proximity is manipulated into policing and disciplining nonnormative sexual desires.

The walk to the wall that abuts the river—a locational mesh of queer bodies—becomes a metaphoric descent to hell for Offred and Ofglen as they traverse a set of stairs from the central city space. The gay man's corpse is a reminder of not only the immediate death an active queer-identifying individual would face in Gilead, but also their eventual fate as transgressive, sinful humans—actual expulsion to hell. Here Gilead's dirty wall—the location of bodily punishment—reduces the gay man to a status physically beneath humanity as he is removed to the side of, and lowered beneath, the cityscape. The gay man's corpse is juxtaposed with those who perform heterosexuality, who are valued and afforded presence on the literal higher ground of the city space. While the association between the queer body and the wetland river space remains a negative ideological construct for Gilead, their similarly constructed identities initiate a confluence between the queer and the natural, which we reclaim to affirm an ambivalent entanglement between natural waters and queer individuals. Although the wall rests on the city's periphery, the handmaids make a point to walk around this river—as Ofglen states, to take the "long way"—which places value on the abjection of space and bodies and subverts Gilead's construction of the river as a hellish, impure locale. The river is perceived as negative because it adheres to an identity specifically "in-between," articulating the environment's transformative abilities, which we suggest as a reason to consider the river as not only a queer space but also one available to trans-identities.[25] In its ability to enact metamorphosis and transformation, water elicits an individual agency and identity independent from human control. Describing such agency and individuality, Rod Giblett posits the wetland as "a place to celebrate . . . ambivalence and fluidity . . . the place of multiple meanings where sense slips and slides never fixing on a point or a definition."[26] The wetland, as a fluid, unfixed entity, never strictly water or land, has a vibrant animacy—as Chen might suggest—that subverts human definition and ownership and promotes similar fluidity, a slipping and sliding away from dominant, normative ontologies of gender, sexuality, and identity. The link, then, between the gay man's body and the riverbed ultimately reveals the potential for transformation and variation within natural water spaces, proving the affinity between queer identities and water and allowing for Ofglen and Offred to recuperate the river space as one that transforms their own relationship into one rich with intimacy, ultimately enabling queer nostalgia and eroticism.

Fluid identities are nostalgically celebrated and desired, in that calling back to the past initiates the preservation and liveliness of sexual identity in the present and future. This preservation—and protest—is initiated when Ofglen and Offred repurpose the abject river into an environment of erotic, fluid sanctuary, despite its echoes of violence or intent as a cautionary symbol. While walking along the river, Ofglen whispers to Offred, "My wife and I had a son" before the construction of Gilead, establishing her queer identity that otherwise is kept hidden. Such a confession constructs an intimate bond between the two women by way of shared personal, sexual histories. The conversation between Ofglen and Offred can take place only at a remove from the city space as they take a walk along the river—the river's own constructed identity as abject and hellish ensures the handmaid's privacy—creating a physical escape and shelter from the city. In claiming this fluid space as a location for the revelation of transgression, the proximity of the river aids and informs their identity and social bond. Neimanis emphasizes the vitality of water for humanity by suggesting that embodiment is a "politics of location, where one's specific situatedness is acknowledged. . . . We require other bodies of other waters (that in turn require other bodies and other waters) to bathe us into being."[27] Through the river site, Ofglen's queer sexuality is (re)bathed into being, as this spatiality allows for her sexuality to be relived and present, creating a momentary feeling of hope and security in this dystopic world.

Female erotic kinship is similarly "bathe[d] into being" as the sharing and acceptance of sexual identities and history near the river fosters an erotic bond between Offred and Ofglen, thereby linking the female body, queer identity, and chaotic waters. This bond results in social anxiety, which is expressed by a male police officer and Aunt Lydia (a figure who similarly polices the handmaids), who interrogate Offred for information once Ofglen's sexuality is publicly revealed. They ask: "Did you ever take the long way home, by the river? On these walks alone, did you ever do anything more than talk? Did she ever touch you? Did she ever try?" These paranoid questions anticipate female sexual intimacy created by and with water, which coincides with "the long way home." The interrogation—Did Oflgen ever touch you by the river?—anticipates a riverine homoeroticism that must be foreclosed by way of punishment. Although Offred and Ofglen do not engage in physical or sexual intimacy, their emotional intimacy fulfills Adrienne Rich's concept of the "lesbian continuum," which includes "a range . . . of women-identified experience; not simply the fact that a woman has had or consciously desired genital sexual experiences with another woman."[28] The intimacy between Offred and Ofglen, situated by the river, enables a subversion of Gilead's patriarchal control: a control that is adamant in preventing the handmaids from any intersubjective sharing and thus stanches semblances of pleasure that accompany subversion and female bonding. Such transformation adheres to Neimanis's description of water as "our buffer, our vital conduit, our solvent, our gestational medium" as the aqueous body here is the gestational force of nonnormative intimacy.[29] The river mediates female, queer, erotic kinship and joy, a feeling Audre Lorde states is fundamental to female erotic relationships. Lorde writes that the erotic is "the power which comes from sharing deeply any pursuit with another person. The sharing of joy, whether physical, emotional, psychic, or intellectual, forms a bridge between the sharers."[30] In extending Lorde, the river space makes possible queer female joy. Without the river to navigate, Offred and Ofglen are denied the possibility of communication, sharing, and community building.

Inasmuch that erotic intimacies constructed by the river are foreclosed, the aquatic nature of the female body itself is similarly repressed as Ofglen is violently punished for her "unnatural" sexuality through clitoridectomy, a bodily violation that forcibly deters the excretion and sharing of sexual fluids by assuring that one cannot achieve orgasm through clitoral stimulation. Such foreclosure is Gilead's means of controlling female sexuality. While femininity has been perceived over time as "wild" and "uncontrollable," the desire to control "uncontrollable" female sexuality is rooted in the female body's wet expression and perceived quality, a nature that induces patriarchal anxiety. While the anxiety over the fluid female body has historically been attributed to the fluids of menstruation,[31] the aquatic nature of female sexuality is also rooted in the orgasm, a violently pleasurable sensation and aquatic intra-action that preoccupies male anxiety over the control of the female body—especially when the orgasm is created by female homoeroticism. Queer female intimacies render the male body useless, as Neimanis similarly notes: "Our watery bodies' challenge to individualism is thus also a challenge to phallogocentrism, the masculinist logic of sharp-edged self-sufficiency."[32] As a source of orgasmic potential through masturbation, the clitoris further represents a subversion of heterosexuality due to its ability to individually create pleasurable fluidity. The violence of Ofglen's cliterodectomy is especially sorrowful given its permanency, as it brutally forecloses both present homoeroticism and queer futurities of pleasure. This futurity is best described by Annamarie Jagose, who engages Dominic Pettman to reason that "the present is the one tense that is anorgasmic: 'orgasm is either future tense (either during the overwhelming presence of the spasm itself), or it is past tense . . . gone, extinguished, history.'"[33] Ofglen is surgically punished for engaging in a sexual affair with an unnamed "Martha"—Gilead's term for female servants—who, importantly, is not her wife, showing a hopeful desire to live her sexuality even after her marriage is destroyed. When Ofglen awakes post surgery, her lower body is clothed in white bandages shaped like a diaper, and she arises from a pure white bed with equally white sheets, found in a similarly and overwhelmingly white room. This whiteness evokes a constructed sterility and cleanliness that Ofglen is now forced into adopting, while her bandages cover and contain her genitals. By covering her genitals, the bandages reflect the suppression of excreting liquid, symbolizing the patriarchal desire to forcibly stanch fluidity and wetness. The violence of hydro-eroticism that loomed while Ofglen and Offred constructed their erotic intimacy near the river is enacted here as violence becomes a means to disrupt and master the wetness of female sexuality and the wet ontology of the female body.

Atwood's novel also maps hydro-eroticism as a prospect of queer sexuality and subsequent violence through the act of showering, located in a private bathroom, a gesture we see recast in *The Shape of Water*. Such a scene is speculated by Offred, who in the novel (alternatively from the adaptation) views three men hanging on the wall, two of whom are queer. Offred imagines their intimacy: "The two . . . have purple placards hung around their necks: Gender Treachery. Their bodies still wear the Guardian uniforms. Caught together, they must have been, but where? A barracks, a shower?"[34] The eroticism of sex is heightened while showering as water becomes an added source of pleasure affiliated with hygiene. Such pleasure derived from water reveals the rich sensuality of hydro-erotic sex, a polyamorous act as water becomes participant, the sex therefore fulfilled by both human and non-human. This imagined yet possible erotic encounter heightens heteronormative, patriarchal anxiety over queer sex due to water's identity as fluid, chaotic, and uncontrollable (calling back to the watery chaos found in Genesis, apt for this fundamentalist society). Water's power to subvert control is rooted in its erratic nature, described by Teresa Shewry, who writes, "Water is temporally and formally unstable and provokes temperamental responses."[35] Shewry and Neimanis engage the concept of "entanglement"; water shapes human responses directly by imbuing the human with water's own characteristics of temperamental tempestuousness. The heightened anxiety over hydro-eroticism also arises as it subverts dominant Christian understandings of baptismal waters as purifying and cleansing. Queer sex engagements by way of hydro-eroticism expose and taint the fallacy of water as infinitely pure and thus dissociated from queerness. Punishment for this imagined act through hanging over the similarly impure river links the queer body and queer sex with aquatic littoral space. The series reflects a perception of the queer body and the river space as unhygienic. Giblett notes that wetland spaces pose "a threat to health and sanity, to the clean and proper body and mind," since wetlands are "neither strictly land nor water."[36] In this way, negative connotations ascribed to wetland spaces similarly reflect the perceived impurity of the queer body, culminating in a threat to social sexual norms. Although Giblett differentiates rivers from wetlands by considering that "rivers flow whereas the waters of wetlands stagnate," we read the littoral space, the river's border—where fluidity and flow are mitigated by interstices of mud and marsh that breed expanses of weeds (plants culturally viewed as undesirably abject)—as wetland. This interstice embodies stagnation and also incurs stagnation of life and sexuality as a site of punishment.[37]

Whereas the shower is rendered a site of queer male intimacy and punishment, the series relays a club bathroom as a space that brings about feelings of melancholy and depression, experiences due to past trauma inflicted on the queer female body. It is the club bathroom that provides yet forecloses eroticism between Moira, Offred's best friend pre-Gilead (played by Samira Wiley) and another woman. Moira, a black woman, resides in Canada after escaping her fate first as a handmaid and later as a sex slave in Jezebel's. In the bathroom Moira and the unnamed woman engage in a frenzied, one-sided act of intimacy, as Moira quickly brings her momentary partner to orgasm. Much as Offred imagines the intimacy between the two gay men showering foreclosed as they are caught and later punished by Guardians, Moira forecloses her own intimacy by refusing her partner's attempts at reciprocating pleasure. Having brought the woman to climax, Moira immediately turns away from the woman and rinses her hands in the sink, cleaning off the fluids her partner has produced. When they finally speak to each other, their interaction is marred by the sound of water slowly dripping from the faucet. Both the camera and the scene's audio magnify this aqueous reminder. Here Moira employs water to cleanse herself from a queer intimacy that she cannot mutually enjoy due to past traumas incurred through forced heterosexual relations (as both handmaid and prostitute). Thus she allows the water to speak the unspeakable: she cannot fully live out her sexual desires as a queer woman because she is trapped in a melancholic state in which she is prohibited from fully healing. While Offred formerly imagines the bathroom as a space where queer intimacies are harbored yet aided in amplified pleasure through aquatic interaction, Moira's bathroom liaison functions as a sorrowful reminder of the need to conceal queerness, the violence Moira previously experienced, and her present inability to fully engage in and enjoy erotic intimacies. Rinsing her fingers, Moira not only limits this intimacy but sets out to ostensibly purify herself, unable to entirely break from her prior life of enforced heterosexuality that demonized queerness.

While the queer body is violently punished due to the potential for wet sexual expression, violence is also enacted against queer bodies directly, as dissenting women in Gilead are punished by deportation to the colonies, where these individuals serve as expendable laborers subject to the dangerous conditions of the postwar environment. A withered wasteland of dirt meshed with desiccated fields, the colony to which Ofglen is remanded following clitoridectomy is a space that slowly kills the female laborers by way of harsh ecological conditions. While the polluted air and land are conditions made apparent through their visibility (the colony looks as if stepped in a dark filter, indicative of the polluted air quality), the condition most harmful to the women is the lack of clean water. While standing over the communal metal sinks in the middle of the women's sleeping area, Ofglen chastises a newcomer for scrubbing her fingernails too roughly with soap and water. The woman states, "I have to get the dirt out, I don't want to get an infection," to which Ofglen replies, "You will. All of the water is contaminated." All of the women are in different stages of varying diseases, suffering from the slow violence of polluted water, resonating with Rob Nixon's articulation of slow violence as "long dyings—the staggered and staggeringly discounted casualties, both human and ecological that result from war's toxic aftermaths or climate change."[38] The effects of slow violence are revealed in bodily damage experienced by the women: some experience hair loss, one graphically removes one of her own fingernails, and Ofglen loses a tooth. These physical modifications represent water's violent potential, which is amplified through a human violence that damages the environment, rendering resources such as clean water obsolete. Just as the queer bodies hanged on the wall in Gilead were ideologically linked to the impure wetland of the river's border, queer bodies and polluted waters are linked in the colonies, as their removal here suggests that queer bodies constitute unclean objects that will be rendered detritus after their inhabitants run their course as useful laborers. The history of viewing queer people as polluted is described by Katie Hogan: "For centuries gay men, lesbians, and transgender individuals have been characterized as 'pollution' threatening the moral fabric of society through a willful creation of dirty, diseased, immoral environments."[39] Polluted waters and the queer body are meshed directly before Ofglen loses her tooth, as she uses her finger to brush her teeth and, feeling discomfort, removes the painful tooth and rinses her mouth out with the brown, murky water of the rusted faucet. Such a loss calls back to her clitoridectomy: Ofglen's body is further brutalized, this time directly due to the slow violence of polluted waters that kill off the degenerate, queer population expelled from Gilead.

Although water—and associations with water—lead to bodily harm for Ofglen, her former positive relationship with water as constructed through the river space is reinvigorated when she returns to Gilead and is reunited with Offred. During this reunion they begin to call one another by their real names (Ofglen is Emily and Offred is June) and resume their walks along the river. On one of these walks Emily reveals that she keeps dreaming about her son, expressing a sorrowful longing to be with him rather than in Gilead; however, she follows this expression by revealing, "I'm glad I got to come back here. I'm glad I got to see you again, June," after which the two hold hands and exchange smiles. This act of touching further bonds the two women and heightens their intimacy as one of lesbian experience, which Rich expands "to embrace many more forms of primary intensity between and among women, including the sharing of a rich inner life" and "the bonding against male tyranny."[40] Since touching between handmaids is strictly forbidden in Gilead, this moment of touching is especially significant in creating a new, physical layer to their intimacy and also in exemplifying a homoerotic experience, as this physical bond subverts Gilead's male tyranny that holds women strangers to one another. In bridging this intimacy through physical touch, Emily and June once more find a brief moment of joy, an essential component of Lorde's conceptualization of lesbian eroticism. While the river here forges intense female bonds, the violence and trauma experienced by Emily due to her contact with polluted water and her fluid sexuality lend a somber quality to this moment. Water proves to home and hone female intimate relations, yet these intimate encounters are both precious and precarious: a harsh realization of queer temporality.

Loving Ourselves and Others in Water

Whereas *The Handmaid's Tale* offers a hydro-eroticism that unveils a state hell-bent on policing, castigating, and effigizing queer individuality and sexuality, *The Shape of Water* engages a hydro-eroticism that lurks beneath an ostensible heteronormativity. Violence and persecution operate on the surface of *The Handmaid's Tale*, forcing queer individuals and nonheteronormative intimacies underground. *The Shape of Water* offers a similar furtiveness to queer interrelationality: the arrangement of a white, female protagonist and a male humanoid creature seems to double down on a (bestial) heteronormativity, which we read as ultimately eschewed. The creature's own Indigenous background (purportedly Amazonian), his aqueous constitution, and the role that water plays in mediating his erotic connections speak to Kohn's articulation of a Runa animism, that is, a belief system that stems from an Amazonian Indigenous perspective on the environment. "Runa animism," Kohn suggests, "grows out of a need to interact with semiotic selves qua selves in all their diversity. It is grounded in an ontological fact: there exists other kinds of thinking selves beyond the human."[41] The hydro-eroticism that *The Shape of Water* yields corresponds with this mode of thinking by way of both the creature and the materiality of water itself. Hydro-eroticism here invites us to immerse ourselves and seek what operates beneath the superficial; it ultimately offers modes of violence that enable the fluorescence of intimacies previously impossible.

Nominated for thirteen Academy Awards and recipient of four, including Best Picture, Guillermo del Toro's *The Shape of Water* (2017) has been hailed by international film critics as an exceptional success. Part fairy tale, part macabre horror, and part monster movie—and wholly cinematographically stunning—*The Shape of Water* envisions a simultaneously futuristic and Cold War–era Baltimore in which a mute custodial worker, Elisa (Sally Hawkins), falls helplessly in love with *The Creature from the Black Lagoon*'s doppelgänger (Doug Jones), who is not given a name and likewise cannot speak. The creature is captured by the US military for alleged magical powers (which the military ostensibly attempts to harness) and subsequent vivisection when the magical procurement is unsuccessful. To free her lover and bring the act of lovemaking to fruition, Elisa is supported by two tokenized characters:[42] Zelda (Octavia Spencer), her black companion and sign language translator, and Giles (Richard Jenkins), her gay best friend and neighbor whose affection is rebuked by a potential flirt and whose only companions, aside from Elisa, are his cats (one of whom the creature decapitates and feeds on). Once the creature is manumitted from the laboratory prison and the relationship consummated—in water, of course—Elisa realizes that their star-crossed love is unsustainable, since he cannot live captive in her bathtub forever. A plan is concocted to free the lover into a channel that leads to the Atlantic Ocean (by way of Baltimore's Inner Harbor) at the height of the rainy season, wherein the fairy tale romance will end and the lover can return to his aqueous abyss.

A. O. Wilson's *New York Times* rave review, "*The Shape of Water* Is Altogether Wonderful," closes with this: "In Mr. del Toro's world . . . reality is the domain of rules and responsibilities, and realism is a crabbed, literal-minded view of things that can be opposed only by the forces of imagination. This will never be a fair or symmetrical fight, and the most important reason to make movies like this one—or, for that matter, to watch them—is to even the odds."[43] The imagination looms immensely in the film—del Toro's masterful cinematography ensures this—but it is coupled with both reality and realism that are, by the film's end, suspended. The suspension of a particular reality, though, does not discredit the film's queer ecological leanings; hydro-eroticism is configured as the sexual liaisons that occur in water and also as the queer unification of the human and nonhuman creature made possible only through immersion and the violation of the body. Throughout the film water is the sole medium wherein sexual intimacy is realized, and water remains the only medium in which sex between two partners is consensual.[44]

Following the film's Best Picture and Best Director wins at the ninetieth Academy Awards, Twitter was aflame with risible reactions, all underscoring the film's most unbelievable suspension of reality: Elisa's congress with her fish lover. User Louis Virtel tweeted, "THE SHAPE OF WATER better be sold overseas as YOU FUCKIN' THAT FISH?," and user Veronica Miller Jamison reported, "I heard somebody refer to 'The Shape of Water' as 'Grinding Nemo' and I'm never going to get over it."[45] These laughable reactions speak to Nicole Seymour's articulation of an "irreverent ecocriticism," which she defines as modes of ecological inquiry that are "absurd, perverse, and humorous in character and/or focused on the absurd, perverse, and humorous as they arise in relationship to ecology and representations thereof."[46] The sexual liaison between Elisa and the creature engenders Seymour's tripartite characterization of irreverent ecocriticism in that it is absurd, perverse, and one of the few moments of comic relief, at least after coitus.[47] The hydro-eroticism of *The Shape of Water* navigates the suspension of disbelief—is she really going to have sex with him? How is that possible?—providing a humorous capacity to eroticism (sex can be fun and funny) that does not undercut experiences of eroticism or a version of queer sex.

The film's opening shot—a dream sequence—is immersive: Elisa's apartment is submerged, and we are led to experience the opening moments "under the sea." Elisa's watery fantasy is broken by the sound of an alarm, which jars her awake and commences her daily routine. But the submerged quality of the dream is responsible for Elisa's arousal, which she manages by slowly and then aggressively masturbating while reclined in a bathtub immediately after waking. The hydro-eroticism of this moment lies not only in the fact that Elisa's daily routine includes this hygienic-masturbatory practice, but also that it calls backward to the fantasy of the dream, which both comforts and exhilarates Elisa, and forward to the consummation of her relationship with her fish lover. The water-filled bathtub—which is noticeably aged and cracked—and later the impossibly flooded bathroom become sites for Elisa to conjure up the masturbatory potential of water, in which water participates as both medium and actor in the autoerotic gratification. The porous nature of the bathtub and the bathroom underscores the fluid intimacy between Elisa and the creature, thus situating aquatic environs as spaces of intimate connection, and, conversely, nonaquatic environs as locations in which intimacy is stanched, foreclosed, or impossible. The film demands that we understand water as the connective tissue that might unite sexual, intimate bonds, especially given that the relationships that exist outside water are doomed: Giles is condemned to loneliness when his flirtation with the male diner-worker is homophobically rejected; Zelda's browbeaten husband proves effete when confronted with the menacing assault of Richard Strickland (Michael Shannon), director of the militaristic laboratory that captured the creature; and Strickland's own perfunctory marriage is consummated only through sadistic sexual violence. The failure of these relationships forces the spotlight on Elisa's relationship with the creature and begs us to consider how water might facilitate a lovingness otherwise absent on land.

Water and the human body share an unmistakable, ineffable coconstitution.[48] To quote Neimanis once more, "Our own embodiment . . . is never really autonomous. Nor is it autochthonous, nor autopoietic; we require other bodies of other waters (that in turn require other bodies and other waters) to bathe us into being."[49] The use of *to bathe* is, of course, poetically figurative, as with the case of Ofglen and queer female sexuality, but Elisa's recurrent bathtub moments allow us to take these words at face value. It is both the fish lover and the medium of water that bathe Elisa into being, resonating with a freedom and intimate reunification that the film's final scene assures we will never forget. Alaimo pinpoints a similar porous blending: "A conception of trans-corporeality . . . traces the material interchanges across human bodies, animal bodies, and the wider material world."[50]

Elisa's engagement with water and her entangled intimacy with her fish lover epitomize the reshuffling and blending of various bodies with unceasing erotic and queer potentiality. Masturbation, an "illicit behavior" that has deep biblical roots starting with Onan in Genesis, is consistently excoriated by the religious right for its tendency to pollute the body and the soul, because it is a sex act that is nongenerative and thus outside the church's view of sex as an act sanctioned only for unification and procreation (of a heterosexual, abled couple). Elisa's quotidian masturbatory practice rejects the procreative model of a heteronormative sexual relation and is further maligned when she begins to fantasize about the creature, who is implicitly understood as male but not human. How humanness can be queer or whether the queer has ever been human are questions that Dana Luciano and Mel Chen take up. Their response to the latter question is both yes and no.

> *Yes*, because this sustained interrogation of the unjust dehumanization of queers insistently, if implicitly, posits the human as standard form, and also because many queer theorists have undeniably privileged the human body and human sexuality as the locus of their analysis. But *no* because queer theory has long been suspicious of the politics of rehabilitation and inclusion to which liberal-humanist values lead, and because "full humanity" has never been the only horizon for queer becoming.[51]

While Elisa and the creature ostensibly operate within a heteronormative nexus, their queer entanglement results from their displacements from corporeal normativity: she is mute, and he does not possess the capability of human language, in addition to his physiology, which is more fish than human despite his bipedalism.[52] For Elisa, hydro-eroticism makes possible myriad intimate connections that nonaqueous environments inhibit. First, elemental water makes possible particular forms of communication, which is of the utmost importance, given both lovers' inabilities to normatively speak. Hydro-eroticism, in this way, attends to how disabled or nonhuman bodies—not identical categories—can be validated and engage in pleasure underwater. Second, Elisa's masturbatory longing for a life submerged suggests that, in the film's case, one shape of water is the concavity of the bathtub wherein aqueous fantasy and self-endowed pleasures abound, a connection that *The Shape of Water* shares with *The Handmaid's Tale*.

But the bathtub is more than home to Elisa's autoeroticism; it is also the site of an impossible sexual congress with a nonhuman creature. Sex between the two conjures up considerations of an off-brand bestiality that rejects a stifling heteronormativity and invites readings of mutually pleasurable experiences between animals and humans. Even more powerfully, it has been fetishized by the film's die-hard fans.[53] First, though, a caveat: by no means do we intend to conflate queer affection or eroticism with bestial affection or eroticism—itself a homophobic, animalizing conflation wielded against queer individuals[54]—and the main difference, as offered by the film, is the issue of consent. Religious and moral considerations aside, bestiality is implicitly prosecuted not only for its perceived lewdness but because it can be read as a form of sexual violence perpetrated against the animal body that in turn maligns the human body.[55] This is not the case in *The Shape of Water*, and thus we bench the conversation of bestiality because the sexual relationship between Elisa and the creature operates under the auspices of consent—reinforcing both with agency and positioning the creature in a realm of reasoning and cognition not usually granted to animals.

Elisa's flirtation with the creature—by way of feeding and nurturing—assumes a hydro-erotic valence. The film positions the hard-boiled egg as a food staple for Elisa, and likewise it becomes an offering for the creature. It is in fact the hard-boiled egg that seals their affection and enables the creature to both trust and fall for Elisa. Though fish species reproduce in a variety of ways—and reproductive sex is neither discussed in nor the point of the film—a common reproductive tactic is that of *ovuliparity*, which does not require insemination by way of vaginal penetration. Female fish species that, like salmon, goldfish, and eels, reproduce via ovuliparity lay eggs and await males of the same species to fertilize those eggs; this entire process is conducted externally. While it is difficult to discern the creature's ichthyo-background, we can read Elisa's offering of the hard-boiled eggs as her suggestion of their sexual compatibility, that is, as a signal of her reproductive capabilities and readiness. Yet it is clear that Elisa and the fish engage in penetrative sex—she describes this to Zelda with humorous reaction—which ostensibly disregards ovuliparity. However, given that the creature is a computer-generated amalgam of nonhuman aquatic species, the process of laying and leaving eggs is one that would be recognizable to ichthyologists and perhaps to the fish lover. The creature, like Elisa, consumes the eggs, situating a lovingness predicated on the sharing of meals and sustenance. While the presentation of the eggs may invoke reproductive capability, the fact that both ingest the eggs engenders a cannibalistic reading (think Madea or Chronos) in which the potential young are consumed. At the same time, the consumption of the hard-boiled egg as a nonreproductive gift situates an additional layer to Elisa and the creature's queer entanglement and rejects the reproductive narrative that otherwise might be imposed, given their ostensible heterosexual arrangement.[56]

As the rainy season progresses and the channel that will become the creature's escape route fills, Elisa hatches her plan to release her lover into Baltimore's Inner Harbor. Such a plan seems to underscore the connectivity of all bodies of water (that one always leads to another, as if a single fluid plane or *mise en abyme*) while disregarding the creature's native Amazonian environs that are, by the film's purview, impossible to return to. Strickland foils the plan by meeting the couple in the industrial littoral space before the creature's release. Giles is injured by Strickland for aiding and abetting the criminals, and the creature is shot several times, potentially imbuing the film with an overwhelming bathos in which the possibility of emancipation is derailed only a body's length from the overfilled channel. But the creature is not mortally wounded; he enacts his revenge by disemboweling Strickland and then joins Elisa in the channel. This moment reinforces an understanding of water as life: it is a home that facilitates the creature's return to an aquatic environment and Elisa's metamorphosis into an amphibious being. Water fosters their intimacy. Moreover, we learn that the orphaned Elisa was found "by the water" and bears gill-like scars on her neck. The camera notices this in the film's opening moments. The creature's own gills and the similarity of these physiological marks on both bodies—human and nonhuman—are unmistakable. Such a marking of Elisa's body, in addition to her speech disability, disrupts normative conceptions of the human and identifies her more closely with the creature, which in this film is not a pejorative identity. Her use of sign language, which she teaches the creature, emphasizes forms of embodied language that make possible communication and connection with nonhuman species. Water enables a life wherein human and creature can communicate without the fetters and cultures of a terrestrial-based mode of being.

The film's ending, like its opening, emphasizes the immersion process, but this time it is not illustrated by a somnolent state. It is in this moment—the one commemorated by the film's cover and promotional materials—that hydro-eroticism reaches its stunning apogee, combining affection, violence, consent, and violation. Submerged in the channel, the creature and Elisa passionately embrace, registering the creature's bioluminescence and healing both of them. But part of Elisa's "healing" is the fact that the creature manipulates her body, transforming her scars into full-fledged gills. The creature's webbed claws, illustrated menacingly throughout the film, here become the means by which physical incision of Elisa's body is the portal to communal being. If hydro-eroticism is deeply attuned to narratives of violence, then in *The Shape of Water* it takes the form of an intimate violence that is both consensual and unifying. This is in direct contrast to the forms of intimate violence that undergird *The Handmaid's Tale* as well as those that Strickland enacts against his wife and threatens to enact against Elisa, fetishizing her inability to speak and thus verbally reject the sexual violence that titillates him. In this way the deployment of hydro-eroticism in *The Shape of Water* reworks the porous boundaries of human and nonhuman bodies that congress and, in so doing, similarly reworks the generative coupling of modes of eroticism with consensual violent engagement, immersion, and antireproduction.

Conclusion

By introducing the neologism of hydro-eroticism, we have attempted in this essay to pull up water's deep feminist roots in order to rhizomatically transplant such a genealogy within the broader discipline of gender and sexuality studies. As our discussion illustrates, hydro-eroticism makes possible forms of queer intimacy and reunion that other terrestrial locations and forms of knowledge inhibit. Given that *The Handmaid's Tale* and *The Shape of Water* are works of speculative fiction that emphasize a necessity for fantasy and the imagination, we suggest that hydro-eroticism anticipates a queer futurity in which violence shadows queer intimacies yet is reclaimed to construct physical, consensual, and unifying bonds. Through bonding waters and (re)imagining intimacies, we have deconstructed concepts of the abject, the inhuman, and the brutally violent that are systematically wielded against queer sexuality to discover and amplify a wet, queer futurity. "Queer bonds," write Joshua J. Weiner and Damon Young, "are what come into view through the isometric tension between queer world-making and world-shattering," and in this vein we offer a similar model of "world-making" that locates water's violent identity alongside queer intimacies in order to situate a future of eroticism, togetherness, and resistance.[57] This togetherness is demanded by hydro-eroticism, a perspective that doubles down on the fluidity of sexuality, the fluidity of human-nonhuman entanglements, and the fluidity of bodies of water that are coconstitutive in the representation of queer kinship. By way of hydro-eroticism, we seek a betweenness that bridges hydro-criticism with gender and sexuality studies in order to underscore the revelatory nature of aqueous, plural intimacies, connections, and embodiments.

JEREMY CHOW is a doctoral candidate in the English Department at the University of California, Santa Barbara. His dissertation examines how bodies of water reorient masculinity through immersive violence in literatures from Defoe to Shelley. His work appears in several edited collections as well as in *Eighteenth-Century Fiction*, *Digital Defoe*, and *Resilience: A Journal of the Environmental Humanities.*

BRANDI BUSHMAN has a BA in English from the University of California, Santa Barbara, with specializations in American cultures and global contexts, literature and the environment, and Native American literature. Her research interests include speculative fiction, gender, sex, sexuality studies, and the environmental humanities.

Notes

1 Alaimo, *Exposed*, 1.
2 Alaimo, *Bodily Natures.*
3 Beauvoir, *Second Sex*, 69.
4 For example, in the original Māori legend on which Disney's very popular *Moana* (2016) is based, Māui climbs inside the vagina of Hine-te-pō, the goddess of death, and as punishment she crushes him with vaginal obsidian teeth.
5 Our use of ontology is different from that of Elizabeth DeLoughrey, who offers "sea ontologies" to consider "the connection between ancestry, history, and non-Western knowledge systems in submarine aesthetics" ("Submarine Futures," 36).
6 DeLoughrey, "Submarine Futures"; Mentz, "Toward a Blue Cultural Studies"; Blum, "Prospect of Oceanic Studies."
7 Barad, *Meeting the Universe Halfway*; Neimanis, *Bodies of Water*, 2.
8 Haraway, *When Species Meet*, 4.
9 Gaard, *Ecofeminism*; Mortimer-Sandilands and Erikson, *Queer Ecologies*; Seymour, *Strange Natures.*
10 Gaard, "Toward a Queer Ecofeminism," 118.
11 Gaard, "Toward a Queer Ecofeminism," 118; Bachelard, *Water and Dreams*; Neimanis, *Bodies of Water*, 84; DeLoughrey, "Submarine Futures"; Giblett, *Postmodern Wetlands.*
12 Gaard, "Toward a Queer Ecofeminism," 118.
13 Chen, *Animacies*, 2.
14 Kohn, *How Forests Think*, 92.
15 Alaimo, "Unmoor," 410.
16 Alaimo, "Unmoor," 416.
17 Atwood, *Handmaid's Tale*, 248.
18 See Foucault on the spectacle of the scaffold in *Discipline and Punish*, 32–73.
19 See Mortimer-Sandilands and Erickson, *Queer Ecologies.*
20 Neimanis, Åsberg, and Hedrén, "Four Problems, Four Directions."
21 We opt for *Shoah* rather than *Holocaust*, given that the latter's use inscribes further anti-Semitism because of its semantic reappropriation. *Holocaust*'s etymology denotes "sacrifice," especially by way of immolation.
22 *OUT* magazine has similarly written on the plight of queer individuals in *The Handmaid's Tale* in comparison to the oppression and persecution of queer individuals in contemporary America under the Trump regime and also during the Shoah. See Brathwaite, "*The Handmaid's Tale* and the History and Future of Queer Oppression."

23 Rachel Stein also studies the river as a transgressive and queer site in "The Place, Promised."

24 Cohen, *Prismatic Ecology*, xxvii.

25 We are invested in trans-discourses alongside hydro-eroticism. Eva Hayward has made a similar gesture in "Transxenoestrogenesis," and Jolene Zigarovich's definition of *trans* as "connoting unstable, transient, or in-between" and "transformation, development, creativity, reorganization, and reconstruction" is also apposite ("Introduction," 4).

26 Giblett, *Postmodern Wetlands*, 20.

27 Neimanis, *Bodies of Water*, 3.

28 Rich, "Compulsory Heterosexuality and Lesbian Existence," 648.

29 Neimanis, *Bodies of Water*, 86.

30 Lorde, "Use of the Erotic," 56.

31 Mentz, "Seep," 287; Giblett, *Postmodern Wetlands*, 148.

32 Neimanis, *Bodies of Water*, 3.

33 Jagose, *Orgasmology*, 209.

34 Atwood, *Handmaid's Tale*, 43.

35 Shewry, *Hope at Sea*, 62.

36 Giblett, *Postmodern Wetlands*, 3.

37 Giblett, *Postmodern Wetlands*, 23.

38 Nixon, *Slow Violence*, 4.

39 Hogan, "Undoing Nature," 241.

40 Rich, "Compulsory Heterosexuality and Lesbian Existence," 649.

41 Kohn, *How Forests Think*, 94.

42 The tokenization of minoritized characters cannot be underestimated, especially in mode with Hollywood's utopic vision to incorporate a broader spectrum of character representations. In this film, especially, critical race and sexuality-studies lenses are vital in assessing this rainbow of representation.

43 Scott, "*The Shape of Water* Is Altogether Wonderful."

44 The only other moment of sex illustrated by the film is between Strickland and his wife, whom he intends to gag—rendering her mute—as some form of sadomasochistic fetish. The wife's consent is questionable. Pointedly, his sexual attraction to Elisa is based in her disability, which renders her mute.

45 Virtel, Twitter post; Jamison, Twitter post.

46 Seymour, "Toward an Irreverent Ecocriticism," 57.

47 The laughter accompanies Elisa's explanation of how she and the creature consummated their relationship, in which Elisa uses hand signals to reveal the creature's capable protractive penis, which remains in a retracted state while unaroused.

48 See Alaimo, "Oceanic Origins"; Helmreich, *Alien Ocean*; Ah-King and Hayward, "Toxic Sexes"; Haraway, *Staying with the Trouble*; and Jue, "Intimate Objectivity."

49 Neimanis, *Bodies of Water*, 3.

50 Alaimo, "States of Suspension," 476.

51 Luciano and Chen, "Has the Queer Ever Been Human?," 188.

52 The film recognizes the creature's home as South America, and he is referred to as a god revered by the Amazonian Indigenous peoples.

53 XenoCat Artifacts, a vendor on kitsch proprietor Etsy, mass-reproduced two versions of a silicone sex toy, "The Asset," that are meant to replicate the fish lover's genitalia. According to *The Wrap*, the dildos sold out even before the telecast of the Academy Awards had finished. See Maglio, "'Shape of Water' Dildo Sales Surge."

54 We need only be reminded of the fearmongering pamphlets and picketing materials—asking whether men would marry animals next because the "sanctity of marriage" had been irreparably damaged—that accompanied the Supreme Court's decision in *Obergefell v. Hodges*, the landmark decision to legalize same-sex marriage in the United States.

55 Kathy Rudy's "LGTBQ . . . Z?" takes up this polemic by way of queer theory and differentiates between bestiality (often conveyed as pornography in which animals are sexual property) and zoophilia, in which animals and humans enter into a mutually beneficial relationship that may incorporate sex. Donna Haraway has similarly written about her interspecies intimacy with her dog, Cayenne, in *The Companion Species Manifesto*.

56 Even more fascinatingly, the film reveals that Elisa's daily routine includes boiling the eggs before her quotidian masturbatory practice. She sets an egg timer for both practices.

57 Weiner and Young, "Queer Bonds," 223–24.

Works Cited

Ah-King, Malin, and Eva Hayward. "Toxic Sexes: Perverting Pollution and Queering Hormone Disruption." *O-Zone: A Journal of Object-Oriented Studies* 1, no. 1 (2013): 2–12.

Alaimo, Stacy. *Bodily Natures: Science, Environment, and Material Self*. Bloomington: Indiana University Press, 2010.

Alaimo, Stacy. *Exposed: Environmental Politics and Pleasures in Posthuman Times*. Minneapolis: University of Minnesota Press, 2016.

Alaimo, Stacy. "Oceanic Origins, Plastic Activism, and New Materialism at Sea." In *Exposed: Environmental Politics and Pleasures in Posthuman Times*, 111–42. Minneapolis: University of Minnesota Press, 2016.

Alaimo, Stacy. "States of Suspension: Trans-corporeality at Sea." *Interdisciplinary Studies in Literature and the Environment* 19, no. 3 (2012): 476–93.

Alaimo, Stacy. "Unmoor." In *Veer Ecology: A Companion for Environmental Thinking*, edited by Jeffrey Jerome Cohen and Lowell Duckert, 407–20. Minneapolis: University of Minnesota Press, 2017.

Atwood, Margaret. *The Handmaid's Tale*. New York: First Anchor Books, 1998.

Bachelard, Gaston. *Water and Dreams: An Essay on the Imagination of Matter*. Dallas, TX: Pegasus Foundation, 1983.

Barad, Karen. *Meeting the Universe Halfway: Quantum Physics and the Entanglement of Matter and Meaning*. Durham, NC: Duke University Press, 2007.

Beauvoir, Simone de. *The Second Sex*. New York: Random House, 1989.

Blum, Hester. "The Prospect of Oceanic Studies." *PMLA* 125, no. 3 (2010): 670–77.

Brathwaite, Les Fabian. "*The Handmaid's Tale* and the History and Future of Queer Oppression." *OUT*, May 5, 2017. www.out.com/television/2017/5/10/handmaids-tale-and-history-future-queer-oppression.

Chen, Mel. *Animacies: Biopolitics, Racial Mattering, and Queer Affect*. Durham, NC: Duke University Press, 2012.

Cohen, Jeffrey Jerome. *Prismatic Ecology: Ecotheory beyond Green*. Minneapolis: University of Minnesota Press, 2013.

DeLoughrey, Elizabeth. "Submarine Futures of the Anthropocene." *Comparative Literature* 69, no. 1 (2017): 32–44.

Foucault, Michel. *Discipline and Punish: The Birth of the Prison*, translated by Alan Sheridan. New York: Vintage, 1977.

Gaard, Greta. *Ecofeminism: Women, Animals, Nature*. Philadelphia: Temple University Press, 1993.

Gaard, Greta. "Toward a Queer Ecofeminism." *Hypatia* 12, no. 1 (1997): 114–37.

Giblett, Rod. *Postmodern Wetlands: Culture, History, and Ecology*. Edinburgh: Edinburgh University Press, 1996.

Haraway, Donna. *The Companion Species Manifesto*. Chicago: Paradigm, 2003.

Haraway, Donna. *Staying with the Trouble: Making Kin in the Chthulucene*. Durham, NC: Duke University Press, 2016.

Haraway, Donna. *When Species Meet*. Minneapolis: University of Minnesota Press, 2008.

Hayward, Eva. "Transxenoestrogenesis." *Transgender Studies Quarterly* 1, nos. 1–2 (2014): 255–58.

Helmreich, Stefan. *Alien Ocean: Anthropological Voyages in Microbial Seas*. Berkeley: University of California Press, 2009.

Hogan, Katie. "Undoing Nature: Coalition Building as Queer Environmentalism." In *Queer Ecologies: Sex, Nature, Politics, Desire*, 231–53. Bloomington: Indiana University Press, 2010.

Jagose, Annamarie. *Orgasmology*. Durham, NC: Duke University Press, 2013.

Jamison, Veronica Miller (@veronicamarche). "I heard somebody refer to "The Shape of Water" as "Grinding Nemo" and I'm never going to get over it." Twitter, March 4, 2018, 6:17 p.m. twitter.com/veronicamarche/status/970468300303069184.

Jue, Melody. "Intimate Objectivity: On Nnedi Okorafor's Oceanic Afrofuturism." *Women's Studies Quarterly* 45, nos. 1–2 (2017): 171–88.

Kohn, Eduardo. *How Forests Think: Toward an Anthropology beyond the Human*. Berkeley: University of California Press, 2013.

Lorde, Audre. "Use of the Erotic: The Erotic as Power." In *Sister Outsider: Essays and Speeches*, 53–59. New York: Crossing, 1984.

Luciano, Dana, and Mel Chen. "Has the Queer Ever Been Human?" *GLQ* 21, nos. 2–3 (2015): 183–207.

Maglio, Tony. "'Shape of Water' Dildo Sales Surge over Oscars Weekend." *The Wrap*, March 6, 2018. www.thewrap.com/shape-of-water-dildo-sex-toy-sales-surge-oscars.

Mentz, Steve. "Seep." In *Veer Ecology*, edited by Jeffrey Jerome Cohen and Lowell Duckert, 282–96. Minneapolis: University of Minnesota Press, 2017.

Mentz, Steve. "Toward a Blue Cultural Studies: The Sea, Maritime Culture, and Early Modern English Literature." *Literature Compass* 6, no. 5 (2009): 997–1013.

Mortimer-Sandilands, Catriona, and Bruce Erickson. *Queer Ecologies: Sex, Nature, Politics, Desire*. Bloomington: Indiana University Press, 2010.

Neimanis, Astrida. *Bodies of Water: Posthuman Feminist Phenomenology*. New York: Bloomsbury Academic, 2017.

Neimanis, Astrida, Cecilia Åsberg, and Johan Hedrén. "Four Problems, Four Directions for Environmental Humanities: Toward Critical Posthumanities for the Anthropocene." *Ethics and the Environment* 20, no. 1 (2015): 67–97.

Nixon, Rob. *Slow Violence and the Environmentalism of the Poor*. Cambridge, MA: Harvard University Press, 2011.

Rich, Adrienne. "Compulsory Heterosexuality and Lesbian Existence." *Signs* 5, no. 4 (1980): 631–60.

Rudy, Kathy. "LGTBQ . . . Z?" *Hypatia* 27, no. 3 (2012): 601–15.

Scott, A. O. "*The Shape of Water* Is Altogether Wonderful." *New York Times*, November 30,

2017. www.nytimes.com/2017/11/30/movies/the-shape-of-water-review-guillermo-del-toro.html.
Seymour, Nicole. *Strange Natures: Futurity, Empathy, and the Queer Ecological Imagination*. Urbana: University of Illinois Press, 2013.
Seymour, Nicole. "Toward an Irreverent Ecocriticism." *Journal of Ecocriticism* 4, no. 2 (2012): 56–71.
Shewry, Teresa. *Hope at Sea: Possible Ecologies in Oceanic Literature*. Minneapolis: University of Minnesota Press, 2015.
Stein, Rachel. "The Place, Promised, That Has Not Yet Been." In *Queer Ecologies: Sex, Nature, Politics, Desire*, edited by Catriona Mortimer-Sandilands and Bruce Erickson, 285–308. Bloomington: Indiana University Press, 2010.
Virtel, Louis (@louisvirtel). "*The Shape of Water* better be sold overseas as *You Fuckin' That Fish?*" Twitter, March 4, 2018, 8:54 p.m. twitter.com/louisvirtel/status/970522859280916480?ref_src=twsrc%5Etfw.
Weiner, Joshua J., and Damon Young. "Queer Bonds." *GLQ* 17, nos. 2–3 (2011): 223–41.
Zigarovich, Jolene. "Introduction." In *Transgothic in Literature and Culture*, 1–22. New York: Routledge, 2018.

Canal Zone Modernism

Cendrars, Walrond, and Stevens at the "Suction Sea"

HARRIS FEINSOD

Abstract This essay is a narrowly drawn exercise in comparison at a narrow passage of marine transit—the Panama Canal Zone. It argues that the spatial typology of the "zone" might supply one of the figures for a tropological history of comparative modernism at sea. The essay follows disparate works—by the Swiss avant-garde poet Blaise Cendrars, the West Indian writers Claude McKay and Eric Walrond, the Nicaraguan poet Ernesto Cardenal, and the American modernist poet Wallace Stevens—into the space of conflicts and disparities that characterizes the Canal Zone as a peculiar choke point of maritime globalization.

Keywords Panama Canal, transoceanic modernism, Eric Walrond, Wallace Stevens, Blaise Cendrars

An illusion of cosmopolitan choice accompanies the reader who contemplates transoceanic modernism's many ports of embarkation—Alexandria, Antwerp, Buenos Aires, Colón, Hamburg, Lisbon, San Francisco, Shanghai, Yokohama, and so on. It is the sort of illusion rhapsodized by Fernando Pessoa's heteronym Álvaro de Campos, the British naval engineer whose poem "Ode marítima" ("Maritime Ode," 1915) praises "O Grande Cais Anterior" (The Great Primordial Wharf) from which he gazes out upon the world.[1] Intoxicated by modernity's quayside coal fumes, Campos expresses in his expansive odes a futurist fantasy of union with the very engines of oceanic passage and the cargoes they carry: "Içam-me em todos os cais. / Giro dentro das hélices de todos os navios" (I'm hoisted up on every dock. I spin in the propellers of every ship). Even Campos's most conditional, reflective lamentations intensify and broaden his yearning for seaborne totality: "Ah não ser eu toda a gente e toda a parte!" (Ah if only I could be all people and all places).[2] In this sort of *saudade*, I cannot help but hear a modernist amplification of the famous Andrew Marvell complaint—"Had we but world enough and time"—that looms large as an epigraph over Erich Auerbach's *Mimesis*, another of the early twentieth century's boldest attempts at a world-literary synthesis. And in hearing this echo, how can I fail to collocate the topic of this cluster of essays—*modernism and the sea*—with the problems and methods of comparative literary history? Such a collocation

57:1, April 2019 DOI 10.1215/00138282-7309721

certainly makes sense if we understand comparative literary history in Auerbach's terms: as a tension among "diverse backgrounds" converging on a "common fate," accessed by a method that seeks out multiple "points of departure" coalescing around a synthetic or "coadunatory" intuition.[3] A reader of transoceanic modernism, balancing the impulses of Pessoa and Auerbach, might well desire to be hoisted up on every dock and to spin in the propellers of every ship—to take in a synthetic view of a cosmopolitan totality made of human cultures fated, by commerce and technology, to connect as never before.

Yet modernist writers themselves often confronted obstructions to the synthetic cosmopolitanism desired or observed by Pessoa and Auerbach, obstructions that took such forms as customs houses, immigration bureaus and passport control, deferred wages, state surveillance, and world war. Thus the phrase *modernism and the sea* refers to a world of literature and art at once connected by intensifying flows and fortified by proliferating blockages. The commercial forms and regulatory regimes of transoceanic technologies like the great liner, the tramp steamer, the commerce raider, the canal tug, and the coastal barge each resynthesize the dialectic between free trade and protectionism, fluid and impeded transit. Literary works situated in the micropolitical environments of such ships, or at their various ports of call, do not follow Auerbach's "coadunatory intuition" so much as they express a version of Aamir Mufti's claim that "world literature, far from being a seamless and traversable space, has in fact been from the beginning a regime of *enforced* mobility and therefore of *immobility* as well."[4] For every Álvaro de Campos, some other character like John Dos Passos's able-bodied seaman Joe Williams shows up "on the beach": undocumented, out of work, and booted by immigration bureaus from La Boca to the British Isles.[5]

In all the coastal nations of the modernizing world, often-discarded works of literature and art—proletarian novels, sentimental plays, avant-garde poems, lyric sea diaries, silent films, radical leaflets, photographs and paintings—attest to this push and pull of connection and blockage and string its tension along several axes of identity and difference. Each work can be understood as a strand in a cat's cradle[6] of crossings between competing sociopolitical discourses of maritime space, including syndicalism and statism, communism and commerce, exile and empire. Some works may go so far as to allegorize the entire system. However, when the archives of modernism at sea are viewed from the perspective of a given national literature or language tradition, these worldly entanglements tend to appear obscure. This may be true even in our contemporary age of transnational and oceanic scholarly turns, for too often an Anglosphere can be mistaken for the world and an ocean for a transactional arena of fluid connection. In recent years I have sought to articulate a different kind of "modernism at sea," whose worldwide imaginaries do not vindicate the usual language of accelerating connections and simultaneities so much as they trace the outlines of missed appointments, deferred arrivals, lags, and collisions. In my view, these negative experiences most often frame the expression of maritime modernism, and any given work tropes this comparative problem of modernist world literature.[7] In doing so, this work's discursive situation tends to implicate other disconnected works from within the maritime world system and to draw in even those works that might refuse the very terms of comparison.

For instance, a story about laboring in the shadows of a great liner will tend to cast in a new light a poem about reveling in a sunny passage on its decks, and vice versa. This essay is a narrowly drawn exercise in such forms of comparison at one narrow passage of marine transit—the Panama Canal Zone. I argue that the spatiotemporal typology of the zone supplies a key figure for a tropological history of comparative modernism at sea. To make this argument, I follow disparate works—by the Swiss avant-garde poet Blaise Cendrars, the West Indian writers Claude McKay and Eric Walrond, the Nicaraguan poet Ernesto Cardenal, and the American modernist poet Wallace Stevens—into the zone of conflicts, disparities, and experiments in sovereignty that best characterizes the Panama Canal as a peculiar choke point of maritime globalization.

The Panama Panic

> . . . the Panama Canal, mechanical toy that Messrs. Roosevelt and Goethals managed to make work when everyone else had failed; a lot of trouble for the inhabitants of the two Americas you have dammed up within your giant locks.
> —John Dos Passos, "Homer of the Transsiberian"

"C'est le crach du Panama que fit de moi un poète!" (It's the Panama panic that made me a poet!) exclaims the speaker of Cendrars's long poem *Le Panama ou Les aventures de mes septs oncles* (*Panama, or the Adventures of My Seven Uncles*, 1918).[8] The "crash" or "panic" in question is the bankruptcy and liquidation in 1888–89 of the French Panama Canal Company, directed by Ferdinand de Lesseps. At that time Cendrars was a two-year-old Swiss boy named Frédéric-Louis Sauser, but it is true that the specter of the crash proved a source of childhood anguish to the young "Freddy," whose father had quit his job teaching math in the watchmaking town of La Chaux-de-Fonds to participate in a rash of increasingly risky and ruinous financial speculations of the sort excoriated by Émile Zola's novel *L'argent* (*Money*, 1891). These included a hotel venture in Heliopolis, Egypt; a German beer export scheme; and an Italian land deal, as well as investments in the canal.[9]

In *Le Panama* Cendrars associates the crash's "importance plus universelle" (universal importance) with the shattering of his generation's comfortable youth. "Car il a bouleversé mon enfance" (it turned my childhood upside down), he recalls:

> Mon pére perdit les 3/4 de sa fortune
> Comme nombre d'honnêtes gens qui perdirent leur argent dans ce crach,
> Mon père
> Moins bête
> Perdait celui des autres[10]
>
> [My father lost three-fourths of his fortune
> Like a number of upright people who lost their money in that crash,
> My father

Less stupid
Lost other people's money][11]

While these events are verifiable elements of Cendrars's notoriously cloudy biography, the incendiary personae of Cendrars's poems rarely speak with a strong commitment to autobiographical veracity. Instead, they voice collective and allegorical aspirations for the project of literature in the age of a shrinking, technomodern world—a world conceived of as canalized and crossed by an ever more routine latticework of steamers, railways, and cables.

Accordingly, we might extrapolate from the exclamation "C'est le crach du Panama que fit de moi un poète!" an underdeveloped origin story for the cascading international avant-gardes of the next several decades. By Cendrars's logic, the avant-garde's poetic vocation was structured by financial speculations of roving state-private infrastructure projects in an era when they began to replace formal empire and attentive colonial administrations with neglectful leases, the wobbly legal regimes of "assigned sovereignty,"[12] the onset of "petromodernity,"[13] precarious migrant labor, the production of civil conflicts,[14] commodity capitalism stitched together by huge communication and transportation networks, and new, extraterritorial zones of free trade. As the architectural theorist Keller Easterling has noted, the Export Processing Zone (EPZ) and the Free Trade Zone (FTZ) that now dominate the spatial form of capitalism were imagined for Panama's Colón as early as 1917, just three years after the first ship transited the completed canal. They were enacted in 1948.[15]

Therefore Cendrars's exclamation suggests an aesthetic modernism at the dawn of "the zone." To be clear, this is neither Mary Pratt's "contact zone" nor Guillaume Apollinaire's Parisian "Zone" but the Canal Zone itself: scion of "historic entrepots and free ports" and predecessor to contemporary EPZs and FTZs, the typologies that, according to Easterling, have rapidly emerged from a backstage "enclave for warehousing and manufacturing" to become the templates of the "world city."[16] If Pratt's model of the contact zone asks us to put together our comparative understanding of global modernism through a focus on the relationships between "travelers and travelees,"[17] the Canal Zone invites us to note the relations of concerns and interests colliding in the teeth of convulsive spatiotemporal and economic logics: booms and busts, annexation and assigned sovereignty, migratory labor and geo-engineering at sublime scales. Hardly the radiant metropolis associated with the historical avant-garde of Apollinaire, the zone is nonetheless the site of an experimental poetry that "beat a rhythm," in Dos Passos's phrase, from an "age of giant machines and scuttleheaded men."[18] The literature of the zone renders visible the collisions of financial, infrastructural, and labor conditions.

Among the avuncular diaspora of Cendrars's *Le Panama*, none of the uncles works the canal per se, but all seven face their fortunes in the shadows of the infrastructural, planetary transformation it represents. These include the "butcher in Galveston" who "disappeared in the cyclone of 1895"; the cabin boy on the tramp steamer who turns into a gold prospector in California and Alaska; and the railroader in India who becomes a Buddhist anarchist, plotting anti-imperial violence against the British in Bombay. Cendrars's uncles might therefore be regarded as

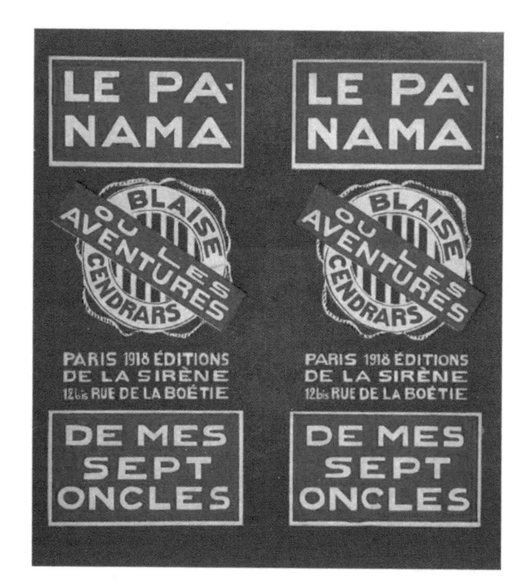

Figure 1. Raoul Dufy, cover art for Cendrars, *Le Panama*.

latter-day "heads" on the "many-headed hydra" that, in struggles with the Herculean forces of capitalism, constitute the master trope of the revolutionary Atlantic for the radical historians Marcus Rediker and Peter Linebaugh.[19] Correspondingly, any "hydro-criticism" worth its salt will probably want to also be a *hydra*-criticism, attentive to the social histories and foreclosed political futures of stateless sailors, deportees, coalers, wharf rats, and other subjects of dispossession forever emerging in the press of neglectful imperial circumstance.

Cendrars portrays his own vanguard company trailing in the wake of these obliterated people on the speculative fringes of capitalist expansion. His poet belongs to a generation of "Jeunes gens / Qui ont subi des ricochets étranges" (Young people / who experienced weird ricochets), who scrounge their way around the globe shipboard in "la cage des méridiens" (the cage of meridians), to quote one of Cendrars's many ways of deflating the glamour of steamer travel.[20] In his own "baptême de la ligne" (baptism of the line)[21] on a 1912 Atlantic crossing more routine than the one he pictures for his nineteenth-century uncles, Cendrars nonetheless imagined that "j'ai partagé tous les sorts du marin. Beau temps des premiers jours, enchantements, vagues, vents, tempêtes, ouragan, dépontement, avaries, dérive, refuge dans un port de fortune. Je n'attendais pas ces choses au XXe" (I shared all the sailor's spells. Initial days of fair weather, enchantments, waves, winds, storms, the hurricane, a swerve, damage, drift, the fortunate port of refuge. I did not expect these things in the 20th C).[22] In the fauvist artist Raoul Dufy's cover image for *Le Panama* (fig. 1), a send-up of a railway timetable, Cendrars's name adorns a life preserver, aligning avant-garde authorship with salvage work in a decade defined by maritime disasters from the *Titanic* to the *Lusitania*.

The Slow Cyclone

Beginning with mid-nineteenth-century projects helmed by competing French and American surveyors to cut diverse trans-isthmian routes through Nicaragua and

Panama, the story of the Canal Zone is well known to us from histories focused on sublime feats of engineering as well as revisionary labor and colonial history.[23] In such accounts, the canal emblematizes the age of the "shrinking world" and what Julio Ramos calls "hemispheric compression"—a swift, commoditized connectivity, often anthropomorphized in boosterist artworks as a "kiss" of the Atlantic and Pacific.[24] Yet, in the framework of transnational and comparative literary history, the shrinking world enabled by the Canal Zone hardly unites the Central American, Caribbean, US, and European vanguardists who variously inhabited it. What would it mean to recompose a literary history of global modernism around the zone, instead of around conventional literary-historical categories of connectivity such as exchange, translation, circulation, intertextuality, and the like? From competing standpoints and historical removes, writers including Cendrars (Switzerland), Stevens (US), Walrond (British Guiana), McKay (Jamaica), Cardenal (Nicaragua), Demetrio Aguilera Malta (Ecuador), Malcolm Lowry (England), Olive Senior (Jamaica), and Juan Gabriel Vásquez (Colombia) all investigate the zone's complex temporalities: sometimes luxuriating in swift cruises to and fro, but more often commenting on its slow, grinding excavations, its swelling locks, its rusting hulks among liana vines, its demolished political horizons and laboring bodies.

In his 1954 poem "Greytown" Cardenal charts this desolate century of speculative development projects that has just passed "como un lento ciclón" (like a slow cyclone) at the great primordial "pier of the Americas."[25] Cardenal's poem offers an anti-imperialist reversal of Ezra Pound's historical montage poetics, ironizing the failures of Cornelius Vanderbilt's Nicaraguan canal projects of the 1850s. At first these projects draw in an exuberantly multilingual, migrant labor community:

> Americanos, alemanes, irlandeses,
> franceses, mulatos, chinos, españoles,
> venían, se encontraban aquí, y partían.
>
> [Americans, Germans, Irishmen,
> Frenchmen, mulattoes, Chinamen, Spaniards,
> they'd all come, meet each other here, and leave.]

But when they rove elsewhere, all they leave behind is one mendicant freebooter who could easily have been one of Cendrars's "uncles": "Edwards E. Brand, de Kentucky, fue el último norteamericano / que se quedó en Greytown, esperando el Canal" (Edwards E. Brand, from Kentucky, was the last North American / who stayed in Greytown, waiting for the canal).[26] Barefoot among the corroded hulls of the paddlewheel steamers that Vanderbilt's concerns have discarded on the shoals of the Mosquito Coast, Brand no longer works as an agent of nascent US imperialism but is fated instead to receive Nicaraguan alms. Cardenal beggars the imperial agencies that arrive at what he calls "the pier of the Americas." Cardenal was therefore uniquely equipped, in 2014, to emerge from retirement for a withering editorial against President Daniel Ortega's concession of a no-bid Nicaraguan canal project to the Chinese Telecom magnate Wang Jing—part of a new "connectography" that makes our age of capricious interoceanic cuts a cynical replay of the first age of liberal globalization that Cardenal had previously decried.[27]

Within Cardenal's slow century, certain synchronic flashpoints make visible the disparate histories of modernism in the zone, such as the decade or so surrounding the Panama Canal's completion in 1914. To put the works of that decade in dialogue with Cendrars's *Le Panama* is to signal the diverse ideologemes the zone represented for Latin American, Francophone, Afro-Caribbean, and Anglo-American modernisms. It renders the Canal Zone not as a feat of connectivity or connectography but as a site of visibility for the conflicting patterns of representation making up a situation of ongoing cultural and ideological disconnection.

The Suction Sea

McKay's dialect poem "Peasants' Ways o' Thinkin'" (1912) suggests the logics by which West Indian migrant labor assembled at the canal.

> We hea' a callin' from Colon,
> We hea' a callin' from Limon,
> Let's quit de t'ankless toil an' fret
> Fe where a better pay we'll get.[28]

The poem's pursuant pro and con calculations are extensive and precise. According to McKay's speaker, the migrant stands to face a regime of legal discrimination and a society that does not recognize Obeah religion. He faces the incalculably bad trade-off of Jamaican rum for Latin American beer and will suffer the loss of sexual gratification, the comfort of family, the experience of village integrity, and more. He will have to learn a new language, do hard labor without weekends off, and face exposure to tropical disease. But face it he will, for the peasant imagines the possibility of remitting money to his family and the promise of a triumphant homecoming rather than permanent diaspora. McKay's peasant can even resign black Jamaicans to a lesser lot than "buccra" (white folk), for like the white of an egg, he announces, "we content wi' de outside."

This kind of resignation seemed implausible to some West Indian writers in the zone. Stories such as "The Wharf Rats" in Walrond's *Tropic Death* (1926), which recollects the world of Colón that Walrond witnessed as the child of a "Panama Man," valorize "the motley crew recruited to dig the Panama Canal" as "artisans from the four ends of the earth."[29] And, even if "dusky peon" imperial subjects of the British, French and Dutch Caribbean supply "the bulk of the actual brawn," Walrond's validation of artisanship over peasant mentality signals an important politicization of Canal Zone labor.[30]

But as far as the Canal Zone is concerned, Walrond also charts more extreme forms of precarity than McKay. In the simple plot of "The Wharf Rats," a West Indian family lives in a shanty by the Colón coaling station. Whenever the tourist ships arrive, two boys—Philip and Earnest—go out in a little rowboat to dive for coins that wealthy passengers toss into the sea. Finally, owing to a convoluted love triangle, a spurned and vindictive woman named Maffi, who practices Obeah, probably curses Philip to his fate: a death by shark attack as he dives in the murky waters for the tourists' coins. The ship for which Philip performs his final dive is named *Kron Prinz Wilhelm*, a ship with an important history (though one that only tangentially includes the Canal Zone).[31] Walrond peoples the ship's deck with wealthy spectators: "Huddled in thick European coats, the passengers" of the ship "viewed from their lofty estate the spectacle of two naked Negro boys peeping up at them from a wiggly *bateau*."[32] Here Walrond offers a stark contrast between West Indian migrant labor and a form of tourism that collapses belle epoque and interwar tropes—more stark even than the protocubist class allegory of Alfred Stieglitz's photograph *The Steerage*. For Philip represents a class that does not access steerage passage. Instead, it earns its living from rickety skiffs in the shadows of the imperial liners, and only when passengers are amused by tossing scraps of multinational currency—American pennies and quarters, British sovereigns, Dutch guilders—into the shark-infested waters. Walrond's description of these waters, as the mixed-up currency sinks, makes it plain how they are figures for the maelstroms of investment and ruination that power the spatial typology of the zone: "It was a suction sea, and down in it Philip plunged. And it was lazy, too, and willful—the water. Ebony-black, it tugged and mocked. Old brass staves—junk dumped there by the retiring French—thick, yawping mud, barrel hoops, tons of obsolete brass, a wealth of slimy steel faced him."[33] The half-submerged ruins of the abandoned project of de Lesseps—where "Iron staves bruised his shins"[34]—prefigure the Middle Passage chains rolling at the bottom of Édouard Glissant's Atlantic, but Walrond's story is shaped less by the history of slavery than by the specific disaster of the "Panama Panic."[35] And here the "yawping mud" deflates the preferred Whitmanian term for voicing New World praise poetry, as if belching a critique of Whitman's celebrations of de Lesseps's Suez Canal in "Passage to India."[36]

A cruel irony presides over the fact that Walrond's attention to de Lessep's discarded remnants, which linger so menacingly as rusted metal in the dark waters of "The Wharf Rats," would only ever be realized as their own fragmentary ruin. Walrond arrived in Paris in 1928 to continue research on a major work of muckraking Canal Zone history, *The Big Ditch*, supported by a Guggenheim grant and an advance from Liveright. He promised to do for the canal what C. L. R. James would do for the Haitian Revolution. In Paris, Walrond doled out praise on Cendrars, noting his particular esteem for *Le Panama*. Liveright advertised *The Big Ditch* in a 1928 catalog, but by 1930 the press had broken the contract for the hundred-thousand-word book. Walrond later serialized pieces of it as "The Second Battle" in *Roundway Review*, the newsletter of the sanatorium where he resided in the 1950s. They are yet to be republished.[37]

Sea Surface

Stevens is among the best-known writers to take the kind of cruise for which Walrond's coin divers performed. He departed from New York on October 18, 1923, aboard the SS *Kroonland* of the Panama Pacific line, calling in Havana, Colón, Tijuana, and finally Los Angeles. His first and only trip through the canal was also his farthest trip south. His wife, Elsie, kept an elliptical diary on the trip, toward the end of which they conceived their daughter. Stevens also conceived a celebrated

Figure 2. The SS *Kroonland* of the Panama Pacific Line passes southward through the Panama Canal–Gaillard Cut on October 25, 1923. Wallace and Elsie Stevens are aboard. James Gordon Steese Family Papers. Courtesy of Dickinson College Archives and Special Collections, Carlisle, Pennsylvania.

poem, "Sea Surface Full of Clouds," venerating the choreography of water and sky off the Tehuantepec peninsula. It was one of Stevens's only poems between *Harmonium* (1923) and *Owl's Clover* (1936). It was also the *Kroonland*'s first trip through the canal since 1915.[38] A photographer captured the ship, newly retrofitted, painted white, and put into use in the trans-isthmian pleasure industry in October 1923, in transit through the Pedro Miguel Lock on the canal, October 25, 1923 (fig. 2). Wallace and Elsie can be imagined promenading somewhere in the frame. That same day Elsie describes the "changeable weather" along the canal and a visit ashore to shop for luxury goods in the free zone, ferried about by a "colored driver" who "spoke English as well as any ordinary darky in Hartford."[39] Observing the casual racism of Elsie's account, it should not surprise us that passengers on Panama Pacific cruises participated in blackface pageantry and cross-class masquerades.[40]

In "Sea Surface Full of Clouds" Stevens's readers have observed a postsymbolist exercise in the production of metaphor through the repetitious, narcissistic descriptions of the "obese machine" of sky and ocean, but in his luxuriant pose of touristic description, outrageous moments of cartoonish racial masquerade interrupt placid reflection:

> And the sea as turquoise-turbaned Sambo, neat
> At tossing saucers—cloudy-conjuring sea?
> *C'était mon esprit bâtard, l'ignominie.*[41]

Walrond's "suction sea" and Stevens's "sea as turquoise-turbaned Sambo" each are forged in racialized revisions of belle epoque literary aspiration, but they do not reflect or embrace each other as do the sea and sky of Stevens's poem. Instead, they begin to suggest how the Canal Zone's compression of hemispheric space enabled only stark consolidations of racial and class divisions. Walrond's Philip dives to his death for the tourists in the same sea Stevens imagines as a reflection of blackface cruise-line pageantry, cut through with the affectations of the French symbolists, whose language Cendrars had sought to jettison.

Another peculiar fact presides over Stevens's placid transits over the "suction sea." He passes over the ruined subaqueous world of de Lesseps that Walrond makes vivid, content to read the water's glimmering surface as the mirror of his inventive acoustical pleasures. But over the next decades Stevens was no stranger to the risky speculations involved in infrastructural projects on the order of the canal, or what he calls the "writhing wheels of this world's business" in his poem "Repetitions of a Young Captain."[42] In fact, Stevens devoted his career to precisely the financial instruments designed to interrupt another "Panama panic." During Stevens's tenure as a bond surety lawyer and later as vice president of the Hartford Accident and Indemnity, the company underwrote construction contracts bonding large-scale transportation and security infrastructure projects, including the Hoover Dam (completed 1931), the San Francisco–Oakland Bay Bridge and the Golden Gate Bridge (completed 1936), the Saint Lawrence Seaway (completed 1959), and the Texas Towers, Cold War radar outposts anchored in the Atlantic for the detection of attacks (completed 1958).[43] But earlier, in the reflective, glittering surfaces of Stevens's poetics, he fashioned distinct bulwarks against Walrond's imagination of the coin diver and its knowledge of the social costs of infrastructural finance.

In all, the financing and construction of an infrastructure space like the Panama Canal Zone—a figure for maritime globalization as the spatial form of capitalism—invite us to reassemble a vision of worldwide modernism structured by disparities among destructive French speculators, the acrobatic West Indians who perform their work among jagged ruins, and the Anglo-American tourist-spectators who take in such performances from aloft. This sort of comparative purchase, from the "wiggly *bateau*" down after the guilders and sovereigns toward the submerged ruins of de Lesseps, and then again up toward the decks of the SS *Kroonland*, might be a model for an account of modernism and the sea that rejects the surface view of Stevens in order to work across divides of class, language, and culture and that remains sensitive to colliding stories of money and labor, cosmopolitan mobility and immobility. Such collisions suggest that a "slow cyclone" or a "suction sea" might be maritime figures that reveal more about how worlds come to connect than do a kiss of the Atlantic and the Pacific or a "sea surface full of clouds."

HARRIS FEINSOD is associate professor of English and comparative literature at Northwestern University. He is author of *The Poetry of the Americas: From Good Neighbors to Countercultures* (2017) and cotranslator of Oliverio Girondo's *Decals: Complete Early Poems* (2018). Recent essays appear in *American Literary History*, *Centro*, *Modernism/modernity*, and *n+1*. His current book project, from which this essay is drawn, is tentatively titled *Into Steam: The Global Imaginaries of Maritime Modernism*.

Acknowledgments

I presented an earlier version of this essay in the seminar "Uneven Development | Capitalist Crisis | Poetics," organized by Margaret Ronda and Ruth Jennison at the 2015 Modernist Studies Association conference in Boston. I gratefully acknowledge the participants in the seminar, and Peter Kalliney in particular.

Notes

1 Pessoa, "Ode marítima," 71; Pessoa, *A Little Larger than the Entire Universe*, 168.
2 Pessoa, "Ode triunfal," 108; Pessoa, *A Little Larger than the Entire Universe*, 160.
3 Auerbach, "Philology and *Weltliteratur*."
4 Mufti, *Forget English!*, 9.
5 Dos Passos, *1919*, 2–42.
6 For an account of the cat's cradle as a figure of comparison, see Morse, *New World Soundings*, 61–89.
7 I argue these points at greater length in Feinsod, "Vehicular Networks"; and Feinsod, "Death Ships." For a welcome turn toward the problem of disconnection, see also Miller and Rogers, "Translation and/as Disconnection."
8 Cendrars, *Le Panama*, n.p. See also Cendrars, *Panama*.
9 Bochner, *Blaise Cendrars*, 16.
10 Cendrars, *Le Panama*, n.p.
11 Cendrars, *Complete Poems*, 34.
12 DuVal, *Cadiz to Cathay*, 31.
13 Whalan, "'Oil Was Trumps'"; LeMenager, *Living Oil*.
14 See, e.g., Mufti, *Civilizing War*, 85–134.
15 Easterling, *Extrastatecraft*, 29–30.
16 Easterling, *Extrastatecraft*, 25.
17 Pratt, *Imperial Eyes*, 7–8.
18 Dos Passos, "Homer of the Transsiberian," 202.
19 Rediker and Linebaugh, *Many-Headed Hydra*, 2–6.
20 Cendrars, *Complete Poems*, 38.
21 Cendrars, *Panama*, n.p.
22 Cendrars, *Inédits secrets*, 178, quoted in Bochner, *Blaise Cendrars*, 27. My translation.
23 McCullough, *Path between the Seas*; Newton, *Silver Men*; Conniff, *Black Labor on a White Canal*; Greene, *Canal Builders*; Senior, *Dying to Better Themselves*.
24 Ramos, "Hemispheric Domains"; Rosenberg, "Transnational Currents."
25 Cardenal, *With Walker*, 78–79.
26 Cardenal, *With Walker*, 78–79.
27 Cardenal, "La monstruosidad del Canal." For a prognostic account of twenty-first-century projects like the Nicaraguan Inter-Oceanic Canal, see also Khanna, *Connectography*. On China's connectographic imaginaries, see Chin, "Invention of the Silk Road, 1877."
28 McKay, *Complete Poems*, 11.
29 Walrond, *Tropic Death*, 67.
30 Owens, "'Hard Reading.'"
31 The actual SS *Kronprinz Wilhelm* never made port near the Canal Zone during its tenure as one of the great prewar German passenger liners. In 1914 the German Imperial Navy requisitioned it as a commerce raider, in which role it captured and sank fifteen British ships off the coast of Brazil in the year before low supplies forced it to dock at then-neutral Newport News, Virginia, where nearly one thousand crew members and officers from the captured ships remained as "guests," building a scrap village called Eitel Wilhelm. The SS *Kronprinz Wilhelm* was rechristened the USS *Von Steuben* and commissioned as a US naval auxiliary vehicle involved in troop transport, leading to its only stop in the Canal Zone to recoal in 1918. The young Walrond might well have seen it on that occasion while working as a journalist for the *Panama Star*, even if it was no longer sailing under the German flag. Walrond's anachronistic placing of the *Kronprinz Wilhelm* in the zone therefore links the relations in maritime history between prewar Euro-American passenger tourism, wartime commerce raiding, and US military involvement.
32 Walrond, *Tropic Death*, 80.
33 Walrond, *Tropic Death*, 82.
34 Walrond, *Tropic Death*, 82.
35 Glissant, *Poetics of Relation*, 6.
36 Whitman, *Leaves of Grass*, 346.
37 Davis, *Eric Walrond*, 325–31.
38 Christened in 1902, the SS *Kroonland* served twelve years, running from Antwerp to New York for the Belgian Red Star Line. Reflagged as an American ship in 1911 for preferential tax purposes, it was put briefly into use as a Canal Zone mail carrier. Repainted in dazzle camouflage, it served as a troop transport ship from 1917 to 1920 and thence returned briefly to the Red Star Line before two years in the Panama Pacific Line. After this it spent two years running from New York to Miami, but the same hurricane of 1926 that Hart Crane experienced on the Isle of Pines destroyed the tourist market the SS *Kroonland* served, and it was scrapped in 1927.
39 Lensing, "Mrs. Wallace Stevens' 'Sea Voyage.'"
40 See the promotional film on the Panama Pacific Line *Over Sapphire Seas* (1934). For a general account of Stevens's lifelong racism, see Galvin, "Race."
41 Stevens, *Collected Poetry and Prose*, 85.
42 Stevens, *Collected Poetry and Prose*, 273.
43 Daniel, *Hartford of Hartford*, 267.

Works Cited

Auerbach, Erich. "Philology and *Weltliteratur*," translated by Maire Said and Edward Said. *Centennial Review* 13, no. 1 (1969): 1–17.

Bochner, Jay. *Blaise Cendrars: Discovery and Re-creation*. Toronto: University of Toronto Press, 1978.

Cardenal, Ernesto. "La monstruosidad del Canal." *La prensa: El diario de los nicaragüenses*, January 11, 2014. www.laprensa.com.ni/2014/11/01/columna-del-dia/216594-dla-monstruosidad-del-canal.

Cardenal, Ernesto. *With Walker in Nicaragua and Other Early Poems (1949–1954)*, translated by Jonathan Cohen. Middletown, CT: Wesleyan University Press, 1984.

Cendrars, Blaise. *Complete Poems*, translated by Ron Padgett. Berkeley: University of California Press, 1992.

Cendrars, Blaise. *Inédits secrets*. Paris: Club Français du Livre, 1969.

Cendrars, Blaise. *Le Panama ou les aventures de mes sept oncles*. Paris: Sirène, 1918.

Cendrars, Blaise. *Panama; or, The Adventures of My Seven Uncles*, translated by John Dos Passos. New York: Harper and Brothers, 1931.

Chin, Tamara. "The Invention of the Silk Road, 1877." *Critical Inquiry* 40, no. 1 (2013): 194–219.

Conniff, Michael L. *Black Labor on a White Canal: Panama, 1904–1981*. Pittsburgh, PA: University of Pittsburgh Press, 1985.

Daniel, Hawthorne. *The Hartford of Hartford: An Insurance Company's Part in a Century and a Half of American History*. New York: Random House, 1960.

Davis, James. *Eric Walrond: A Life in the Harlem Renaissance and the Transatlantic Caribbean*. New York: Columbia University Press, 2015.

Dos Passos, John. "Homer of the Transsiberian." *Saturday Review of Literature*, October 16, 1926.

Dos Passos, John. *1919*. New York: Harcourt, Brace, 1932.

DuVal, Commander Miles P., Jr. *Cadiz to Cathay: The Story of the Long Struggle for a Waterway across the American Isthmus*. Stanford, CA: Stanford University Press, 1940.

Easterling, Keller. *Extrastatecraft: The Power of Infrastructure Space*. New York: Verso, 2014.

Feinsod, Harris. "Death Ships: The Cruel Translations of the Interwar Maritime Novel." *Modernism/modernity*, Print Plus, 3, cycle 3 (2018). doi.org/10.26597/mod.0063.

Feinsod, Harris. "Vehicular Networks and the Modernist Seaways: Crane, Lorca, Novo, Hughes." *American Literary History* 27, no. 4 (2015): 683–716.

Galvin, Rachel. "Race." In *Wallace Stevens in Context*, edited by Glenn McLeod, 286–96. Cambridge: Cambridge University Press, 2016.

Glissant, Édouard. *Poetics of Relation*, translated by Betsy Wing. Ann Arbor: University of Michigan Press, 1997.

Greene, Julie. *The Canal Builders: Making America's Empire at the Panama Canal*. New York: Penguin, 2009.

Khanna, Parag. *Connectography: Mapping the Future of Global Civilization*. New York: Random House, 2016.

LeMenager, Stephanie. *Living Oil: Petroleum Culture in the American Century*. Oxford: Oxford University Press, 2014.

Lensing, George. "Mrs. Wallace Stevens' 'Sea Voyage' and 'Sea Surface Full of Clouds.'" *American Poetry* 3, no. 3 (1986): 76–84.

McCullough, David. *The Path between the Seas: The Creation of the Panama Canal, 1870–1914*. New York: Simon and Schuster, 1977.

McKay, Claude. *Complete Poems*, edited by William J. Maxwell. Urbana: University of Illinois Press, 2004.

Miller, Joshua L., and Gayle Rogers. "Translation and/as Disconnection." *Modernism/modernity*, Print Plus, 3, cycle 3 (2018). doi.org/10.26597/mod.0062.

Morse, Richard. *New World Soundings: Culture and Ideology in the Americas*. Baltimore, MD: Johns Hopkins University Press, 1989.

Mufti, Aamir. *Forget English! Orientalisms and World Literatures*. Cambridge, MA: Harvard University Press, 2016.

Mufti, Nasser. *Civilizing War: Imperial Politics and the Poetics of National Rupture*. Evanston, IL: Northwestern University Press, 2017.

Newton, Velma. *The Silver Men: West Indian Labour Migration to Panama, 1850–1914*. Kingston: Institute of Social and Economic Research, University of the West Indies, 1984.

Owens, Imani D. "'Hard Reading': US Empire and Black Modernist Aesthetics in Eric Walrond's *Tropic Death*." *MELUS* 41, no. 4 (2016): 96–115.

Pessoa, Fernando. *A Little Larger than the Entire Universe: Selected Poems*, translated by Richard Zenith. New York: Penguin, 2006.

Pessoa, Fernando [Álvaro de Campos]. "Ode marítima." In *Orpheu 2*, 71. Lisbon: Ática, 1915.

Pessoa, Fernando [Álvaro de Campos]. "Ode triunfal." In *Orpheu 1*, 108. Lisbon: Ática, 1915.

Pratt, Mary Louise. *Imperial Eyes: Travel Writing and Transculturation*. 2nd ed. New York: Routledge, 2008.

Ramos, Julio. "Hemispheric Domains: 1898 and the Origins of Latin Americanism." *Journal of*

Latin American Cultural Studies 10, no. 3 (2001): 237–51.

Rediker, Marcus, and Peter Linebaugh. *The Many-Headed Hydra: Sailors, Slaves, Commoners, and the Hidden History of the Revolutionary Atlantic*. Boston: Beacon, 2000.

Rosenberg, Emily S. "Transnational Currents in a Shrinking World." In *A History of the World*. Vol. 5 of *A World Connecting, 1870–1945*, edited by Emily S. Rosenberg, 815–998. Cambridge, MA: Belknap Press of Harvard University Press, 2012.

Senior, Olive. *Dying to Better Themselves: West Indians and the Building of the Panama Canal*. Kingston: University of the West Indies Press, 2014.

Stevens, Wallace. *Collected Poetry and Prose*. New York: Library of America, 1997.

Walrond, Eric. *Tropic Death*. New York: Liveright, 2013.

Whalan, Mark. "'Oil Was Trumps': John Dos Passos' *U.S.A.*, World War I, and the Growth of the Petromodern State." *American Literary History* 29, no. 3 (2017): 474–98.

Whitman, Walt. *Leaves of Grass and Other Writings*, edited by Michael Moon. New York: Norton, 2002.

Maritime Optics in Sea-War Fiction between the Wars

NICOLE RIZZUTO

Abstract Absent from prodigious critical scholarship about the seas is a discussion of modernism between the wars. Yet this period is rife with writing set on the waters. This essay argues that it is in their orchestrations of the waters as dead zones that such works revitalize seafaring literature and broaden our understandings of modernism generally. By way of illustration the essay examines the British author James Hanley's 1938 novel *Hollow Sea*, which centers on a merchant ship turned troopship during World War I. In its staging of maritime technologies and infrastructures, Hanley's text ironizes the literary trope and geopolitical concept of the "free sea" from an interwar perspective. The novel's particular mode of hydro-criticism manifests in its formal challenges to both the war optics of the British state and the optics of a major modernist writer of the seas, Joseph Conrad.
Keywords hydro-criticism, James Hanley, modernism, maritime infrastructure, World War I

The interwar period appears to be a dead zone in the history of seafaring literature, a genre that has animated literary and cultural studies for decades, from eighteenth- and nineteenth-century studies to genre studies to postcolonial studies.[1] Absent from prodigious scholarship about the seas is discussion of modernism between the wars, even from otherwise comparative, comprehensive analyses that span the early modern period through today.[2] Moreover, there has been to date no sustained study of maritime modernism; thus one might easily conclude that by the 1920s and 1930s the seas had ceased to serve as a site of literary inspiration, invention, or even representation. Sea fiction appears to die when Joseph Conrad's narratives abandon the waters for South America's silver mines and Europe's metropoles, and is resuscitated only after World War II.

This is not the case, however. If we replace "strong," accepted definitions of modernism based on select formal devices and canonical authors with more flexible approaches attuned to micro- and macrohistories, ecologies, and modes of power-knowledge peculiar to the waters, we discover a vibrant, if submerged, archive of maritime modernism. The sea is the center of numerous interwar writings largely

ENGLISH LANGUAGE NOTES
57:1, April 2019 DOI 10.1215/00138282-7309732

forgotten or neglected today. I contend that it is in their orchestration of the waters as sites of intercepted and failed trajectories that these works of fiction quicken and revitalize seafaring literature in ways that texture and broaden our understandings of this genre, as well as modernism generally. Rather than theaters of freedom, adventure, or progress—a dominant critical interpretation of the waters during this era of unprecedented transoceanic travel—the seas, as portrayed by writers who identified as not tourists or passengers but working-class mariners, are theaters of imprisonment and deflected and dilatory motion resulting from exigencies of the age.[3] These exigencies include a rebalancing of global power in the aftermath of one world war and preparations for another; the decline of imperial modernity's heretofore most powerful maritime empire, Britain's, and the concomitant rise of other sea powers; and an epistemic shift in perspectives of aqueous space created by both historical trends. This shift is a visual remapping of the surface and depths of the seas coincident with the development and deployment of communication and military technologies along with new geopolitical strategies of surveillance and information gathering. Interwar writing registers the restructuring of global circuits of power and attendant transformations of ways of seeing by presenting the waters as cramped with networks that blindside travelers and stymie movement. In doing so, it reworks tactics that appear throughout sea narratives from the seventeenth century through the high modernist era, a slice of modernity during which, as Carl Schmitt put it, "England was strong enough to usher in a new balance [of land and sea] which impeded sea powers, and thereby allowed her alone to dominate the great expanses of the world's oceans."[4]

To support these claims, I turn to a novel that confronts this period in maritime modernity by revisiting World War I through the post- and prewar lenses of 1938: the British author James Hanley's *Hollow Sea*, which narrates the thwarted trajectory of a merchant ship commandeered by the British Admiralty. Like much forgotten sea fiction of the interwar era, Hanley's writings have been categorized as working-class realism. Yet in depicting the war of 1914–18, *Hollow Sea* stages a defining event of the modernist era and contouring force of modernist literature. What differentiates Hanley's allegedly realist war novel from much territorial-oriented modernisms is that the former's textual strategies are shaped by the geopolitics of aqueous extraterritoriality and focalized on, and through, laboring mariners on the high seas, rather than on the consciousness of middle-class characters or returned soldiers in the metropolis. But these strategies also erode boundaries between modernism and realism while supplementing terrestrial modernist perspectives *and* those of literature shaped by maritime history, most notably Conrad's. Conrad's narrative techniques, whether employed to tell stories that take place on sea or on land, are structured by the hydrosphere's systems. Like Conrad, Hanley employs impressionism, foregrounding the struggle to acquire a full and accurate picture of reality, and he, too, showcases the imagination formed by nautical practices. But Conrad calls on problem-solving techniques that emerge from handling logistical tools, navigational infrastructures, and the networks they serve—namely, those that facilitated the expansion and consolidation of Britain's maritime empire—in order to make the abstract concrete and allow narrators and characters to pass through impasses, physical and metaphysical, and clarify vision. Hanley, on the contrary, considers how such tools, infrastructures, and networks have been harnessed to secure the imperial-national power of a state whose maritime might is waning and under threat, and he elaborates how they block sight, abstract space, and confine and stall the passage of those subject to its hydrarchy. His novel's critique of wartime optics is enacted through formal tactics that depart from the protocols of documentary realism. What animates them is not an aesthetic commitment to experimentation, as is the case in certain forms of modernism, but the experiences of those compelled to cross seas made deadly by modern necropolitics, and the strategies these precarious subjects developed to respond to them.

Hanley wrote many works about life at sea from the perspective of merchant seamen like his younger self; *Hollow Sea*, which details the botched mission of a merchant ship ordered to transport fifteen hundred troops to the Dardanelles, is an example. Hanley himself served on merchant vessels turned warships that transported soldiers during this disastrous campaign. In focusing on the Great War at sea, an underrepresented and understudied event in literary and cultural history, this novel forms both a complement and a counterpoint to modernist depictions of the war composed from the perspectives of land and air. As Patrick Deer and Paul Saint-Amour argue, modernist forms exposed and challenged the British state's aspirations to acquire total visual control of these theaters of war. Deer shows that propaganda that touted the state's panoramic vision of all fronts was undermined in poetry and narratives that related the actual experiences on the ground of its "oversights," the compartmentalized and partial vision induced by the labyrinthine trench system and the army's chain of command.[5] Saint-Amour demonstrates that the Royal Air Force's confidence that aerial photography afforded unfettered access to what was camouflaged below through the camera lens's vertical penetration of surface topographies from high above was complicated by the interpretive demands involved in decoding these images—images that artists such as Picasso found to possess disturbing formal affinities with their own works.[6] If the war manifests in canonical modernism as traumatic flashbacks in the consciousness of soldiers returned home from the claustrophobic trenches at the front, as in the works of Virginia Woolf, Rebecca West, F. Scott Fitzgerald, and Ford Maddox Ford, and if it resonates within the rigid visual geometries of cubist paintings that resemble a flattened landscape viewed from the skies, it registers in *Hollow Sea* through the impressions of civilians suspended by the state in a fluid medium that erodes discernible boundaries between war zone and safe zone, and whose vision struggles to penetrate its surface and depths in order to move through it.

This framing of the seas as opaque and deadly intervenes in a dominant, popular imagination of them as ever more lucid and tamed. In many ways, the waters *had* been rendered progressively legible and navigable during the previous half century, as hydrographers began to chart the oceans' depths and naturalists to analyze its inhabitants, undersea cables were laid across ocean floors, the hydrophone came into use, and the steamship accelerated and exponentially increased leisure and commercial travel. *Hollow Sea*, however, insists that the seas had also become ever more obfuscating and impassable under geopolitical changes and technoscientific developments during the previous decades. By articulating a discrepancy between the state's panoptic surveillance of the aqueous globe from the vantage point of central command on land and the compromised sight of its subjects from the vantage point of the waters, the novel ironizes Britain's wartime optics. Specifically, it indicates how tools, infrastructures, and techniques once used to open the seas to make them free for movement are haunted by the aims of imperial world making; as such, they have become imprisoning, confining, and disorienting to those who serve the war effort. This indication often occurs through intertextual nods to maritime literary culture.

To portray a deflected voyage through the eyes and voices of working-class seamen, Hanley draws on a tradition of sea writing while challenging some of its most cherished tropes and ideals, along with the formal practices of its great modernist exemplar, Conrad. During the centuries in which Britain came to rule the waves, sea stories across genres cathected its geopolitical aspirations in portrayals of the waters as boundless spaces of adventure and peril. Voyages were regularly stalled and threatened as a matter of narrative course, whether by the extreme elements; by pirates, sea bandits codified as "enemies of humanity" by international law; or by other unruly subjects who would thwart the interests of the nation-state and the empire embodied in the ship. Such topoi recur with variations across time, in Shakespeare's *The Tempest* (ca. 1611) and Defoe's *Robinson Crusoe* (1719), Coleridge's "Rime of the Ancient Mariner" (1798) and Stevenson's *Treasure Island* (1883), all the way through Conrad's *The Nigger of the Narcissus* (1897), *Typhoon* (1902), and *The Shadow-Line* (1916). Parsing different lineages of the sea-writing tradition, Margaret Cohen analyzes how "craft," a set of diverse skills developed in the age of sail, provides a narrative grammar for writers to deal with impasses across centuries. Although in historical terms craft declines in the age of steam, in literary terms it survives in the formal strategies of modernists such as Conrad.[7] Cohen argues that Conrad's most famous narrator, Marlow, "deploys techniques that he has learned in the course of the mariner's work of navigation, achieving orientation from partial information" and that he does so both on sea and on land. An imperial maritime ethos and practice thus informs more than Conrad's sea narratives, narrowly defined; it also shapes the procedures of such texts as *Lord Jim* and *Heart of Darkness*, the work that *Hollow Sea* references more than any of his others.[8] That such references occur in a novel about intercepted passage that problematizes orientation would seem to align it with Conrad's modernism and the literary genealogy described above. But in Hanley's novel, what disrupts progress and compromises vision are not the enemies of imperial nation-states but those states themselves. The waters are not boundless but captured by networks of mines, submarines, destroyers, hospital ships, and merchant ships, whose presence is ratified by the bureaucracy of nation-states, international law in the form of the Hague Conventions of 1899 and 1907. And rather than identifying with Marlow's perspectives, the text redirects them to point out the effects of the manipulation of the aqueous world to make it navigable for imperial power: a constraining and rendering vulnerable those who labor under its control. *Hollow Sea* suggests that in an effort to totalize its viewing capacity and make the waters transparent, the British war state has obscured them, leaving those on the seas in a nebulous space where they remain subject to its authority but outside its protection.

Both the novel's structure and its rhetoric convey that the Admiralty commands its military and merchant fleets with an exercise of power that is mystifying and impedes navigation. *Hollow Sea* withholds the ship's trajectory and destination from captain, crew, soldiers, and readers in its initial chapters and again later, when the *A.10* awaits orders of where to proceed now that it carries dead and dying troops and crew shot attempting to disembark at Oran. It redoubles the charge that the state disorients the seas, obstructs passage across them, and makes them deadly by deploying an impressionistic rhetoric evocative of Conrad's depiction of a misty, unfathomable Congo viewed through Marlow's colonizing haze. The perlocutionary act that converts a merchant vessel under articles of war is an incantation that kills, transforming the vessel into a death ship: "Her name had gone and the spell was cast. In those few hours it seemed as though the old life in her had been cast out. Nothing remained now but the bare bones of a future . . . uncertain and invisible. It lay hidden in the unfathomable depths of the mist that hung like a pall over the living waters of the seven seas."[9] This passage suggests that what is "unfathomable" is not an Africa that lacks the order of Europe, as Marlow's impressionism indicates. Unfathomable are high seas that European powers have disordered and made unclear.

The text also references *Heart of Darkness* through a figure whose function is purely symbolic in *Hollow Sea*, a mysterious, disheveled character outside the ship's hydrarchy. The story points to the limits of this focalizer's impressions, implying that they are residual, outmoded. The "Rajah" is a squatter who pontificates to the crew from a box and whom the captain orders removed before they set off. He possesses a "bald head that shone like dull ivory," and the captain "imagined him to be a sort of figure-head. A figure-head for a mad ship" (6), later musing that he "could have nailed him under the bow, arms outstretched. He would have plugged the jaws. They irritated him so" (22). Connecting the ship's itinerary to Kurtz's ill-fated journey in the Congo, this figurehead prophesies the *A.10*'s destiny, naming it "a death ship" heading toward unimaginable horrors (232). By projecting Kurtz's vision of Africa onto the high seas, the text insinuates that both are "free spaces," subject only to the law of the strongest. The reference to Conrad's work is again ironic, however. Even by the time *Heart of Darkness* was written, neither space possessed the status of free, that is, operating "beyond amity lines," subject to conquest where only the law of the strongest rules.[10] More to the point, *Hollow Sea* insists, the sea is no longer free because the state has extended its purview over the open waters through the manipulation of tools and techniques that make it impassable.

Hanley's novel reworks the classic narrateme of the sea as a wide-open space and escape from authority by generalizing and recasting a figure likewise drawn from sea fiction—the net. This term at first conjures precapitalist and perhaps premodern sea labor, summoning an image of fishermen on trawlers hauling in their catch. *Hollow Sea* resignifies it through an interwar critique of the instruments of navigation used to graph, calculate, and measure the waters to make them crossable. The captain of the merchant vessel turned death ship, Dunford, warns the crew, "It's going to be pretty hard going before we get through this net" (55). We learn that by "net" the captain is "thinking of the orders, of the official chart. Dangerous and advantageous points, currents, observations and wind, soundings. . . . He must only go by a certain route. At a certain time . . . the chart his infallible guide. . . . He knew every inch of the ground as well as his own hand, but the official voice announced: 'This route and no other' . . . the mathematical calculations of lunatics" (55–56). Normed by the Admiralty as an accurate, neutral reflection of aqueous space enabling ships to navigate, the chart as net is characterized by Dunford contrarily as both simulacrum and trap, obstructing instead of facilitating passage. It is "a world, a wall, a prison, a well" (3).

The novel develops this portrayal of the net by depicting the chart as an unstable representational form that confuses navigation and does not accurately reflect an existing world but instead generates an alternative, fragmented one through abstraction. Challenging the Admiralty's vision and ordering of space by redirecting Conradian impressionism again, *Hollow Sea* opens with a chart-reading scene that recalls the map-reading scene in *Heart of Darkness*. In that episode, imperial powers' redrawing of space is viewed by Marlow as an ordering of the earth that distinguishes beneficial and beneficent exercises of European control from poor and malevolent ones. The map in the company office graphically illustrates in distinct colors the differences between one form of colonial rule, the "good work" done in British colonies, and another form, the questionable activities in the Belgian Congo.[11] Hanley turns Marlow's perspective on its head when he begins the novel by having his merchant captain stare at the map that directs his own incipient journey. He characterizes the nautical chart, an allegedly neutral, scientific logistical instrument, as a geopolitical instrument: the world map drawn through imperial contests for global power.

> His finger now covered the whole area of the chart. The paper itself was covered with lines; red and green and blue. . . . He sat back. When he looked at the chart again it was a mere blur. The colours had run into one another, the lines moved. He put his finger down on the paper. He thought it had moved. . . . Maybe it was just the wind. Then he rolled up the chart, put it in the drawer, and remained standing there, staring at the drawer into which he had placed it. (2)

Viewed through Dunford's eyes, the divisions that organize global space are not stable. That colors "blur" into each other compromises the moral distinctions between forms of rule Marlow sees them as indexing. If the "lines" of latitude and longitude "move," then the inscriptions that coordinate time and space with the imperial center of Greenwich as their anchor are not fixed; navigation across the seas they should facilitate is therefore also threatened. The chart's behavior suggests that it has a life of its own, that it might have "moved" of its own volition, which disquiets the captain and challenges his mastery as he attempts to stabilize it with the emblem of manual labor, his hand. Its techniques of world making, though aspiring to a total, godlike view, abstract and even delete or cross out elemental and human worlds under its purview. As such, the chart is obfuscating and irrational. When Dunford looks at a point on the official chart marked X, "he knew it was really land, sky, water, air, people living and moving about; a place on the earth's surface, but it was called X. He thought: 'Perhaps conundrums are the square root of their crazy philosophy'" (3–4). The net, here the chart, is likened to a magnet, or a compass, because it tends toward totalization in order to provide direction—"it was like a powerful magnet . . . it drew everything unto itself"—but it remains partial, incomplete in crucial ways: "everything but the unrest" (3).

The compass is another metonymic displacement of the net that crystallizes the politics of seeing at sea. Under modern conditions of warfare, the compass is not neutral, or a navigational aid, so much as coercive and confining. It also draws everything to it and imprisons the seas, rather than opening them to make them free for movement. Moreover, it subjects the scopic practices of the seamen to its control. After the ship must proceed without orders following its aborted mission at Oran, the quartermaster steers with his "eyes to the compass, his world prisoner of that world. Beyond, sky, space, water, passing craft, wreckage, a great liner, a rowing-boat, a destroyer, a huge battle-cruiser. He saw only his needle-point" (232). When the chart leads the ship into waters infested with U-boats, drawing them into a spiraling vortex as it attempts to outmaneuver one, the compass does not orient but traumatizes or "pains" vision and disorients the crew: "The quartermaster closed his eyes for a moment. The compass dizzied him—sometimes he saw it spinning as though she were swinging free, as though the spirit of *A.10* broke clear, stood outside the control of men. . . . The compass-point pained the eye, it was the edge of a dizzy whirling world" (103). We might read this episode as a resignification, from a peculiarly maritime perspective, of the wider modernist characterization of the modern world as chaos, as spinning out of control. This passage also summarizes the impasse the novel continually articulates: even in the so-called free seas, the ship's "spirit" cannot stand "outside the control of men," or break free from authority, because it is trapped by nautical infrastructures operating in the service of the war state. This net threatens the lives of those gathered under its legal protection because it both obscures and has an obscured vision of the seas. It contains holes.

The significance of the figure of the net, while metaphorical, is also historical. It indexes a modern shift in sea-war strategy. This is the initiation of network-centric warfare, which began during World War I and continues today. This form of warfare relies on a restructuring of the spatiotemporal relationship between land and sea to achieve panoptic control of aqueous space. As one military historian explains, "Network-centric warfare is really picture-centric warfare, a kind of warfare using a more-or-less real-time picture of what is happening."[12] The inventor of this form of naval intelligence, which dates back to the first decade of the twentieth century, is the British admiral John Fisher, who developed this strategy to obtain "the 'all-seeing eye' of his fleet."[13] First implemented during the Great War, it was controversial—"hated," even. It relied on ships at sea transmitting their own particular information to central command in London, where it would theoretically accumulate to form a full, synchronic picture of the situation on the waters. Pieces of that information would then be transmitted back to ships to guide them. In theory, the Admiralty on land, as opposed to ships at sea, possessed an "all-seeing eye." In actuality, it did not possess as clear or complete a picture as it needed, and nowhere near as complete as that toward which Fisher aspired. Unaware of this limited vision, ships continually failed to act and to communicate coordinates that would fill in the picture. They assumed that they should await orders, which often did not come. Worse, there were more merchant ships at sea than ever before, and real-time control over such highly populated seas proved impossible. In addition to these issues, communications between the army and the navy were abysmal. What has been called "'the curious gap between the mental worlds of the navy and army' . . . extended to misunderstandings over the definitions of terminology as basic as the time of day."[14]

Writing from a postwar perspective that coincides with the increasing threat of a more catastrophic war than the first, fought with deadlier technologies, Hanley emphasizes that traumatic events result from the hubris of the Admiralty's imagining that it could seamlessly coordinate and synchronize the movements of the clogged seas through its terrestrial-based perspective and deployment of logistical technologies. That the time-spaces of land and sea are out of joint is made clear in the way the botched mission at Oran is portrayed. The *A.10* has been ordered to drop the men not only under the light of a full moon but at the wrong beach, at the wrong tidal point, where they are visible, vulnerable, easily shot and killed. The state's failure to paint a picture on land of the theater at sea is attributed to its failure to use the practical knowledge and techniques of the seamen. As Dunford relates, "The army says this, the navy thinks that—as for the ethereal voice, what it mumbles is incomprehensible" (236). Clarifying that the ethereal voice refers to "the power behind all this," he recounts that "the boats full are lowered, are rowed away, and fire is poured upon them. . . . There is no order, no sense of direction, the distant voice is strangely silent" (237). The Admiralty's false impressions of time and space reveal the failure to obtain a picture of the seas in real time—not merely of military operations but of such elemental conditions as the tide and position of the planets, by which sailors have oriented themselves effectively for centuries. In the aftermath of these events, the captain recapitulates his indictment of the state for turning the seas deadly: "One was dragged into the filth with the others, soiled. One didn't even believe in their precious conundrums, nor in those far-off voices that not only mapped lands and oceans, but poisoned the wells of living there" (453). This portrayal of deadly seas highlights the distinctions between Hanley's work and novels that preceded it.

By making the state the cause behind obscured impressions and intercepted passages, *Hollow Sea* distinguishes itself from sea writing that attributes these causes to the elements, to encounters with others attached to colonial spaces, or to both. Conrad's works again come to mind, not only *Heart of Darkness*, where because of delayed decoding Marlow's boat is ambushed by Congolese hiding along the misty river, but also those narratives narrowly classified as "seafaring." In *Typhoon* weather in the China seas assaults the ship and the captain's eyes: "He was trying to see, with that watchful manner of a seaman who stares into the wind's eye as if into the eye of an adversary. . . . Stricken by a blind man's helplessness," he fears "his sight might be destroyed in the immense fury of the elements."[15] In *The Shadow-Line*, when a ship stalled in a windless Gulf of Siam leads to deaths not by enemy fire but by fever, these events are mistakenly attributed to human causes. They stem from the previous captain's obsession with a "low class" white woman, a "medium" or "fortune teller" in Haiphong "disguised in some semi-oriental, vulgar . . . costume."[16] The present captain's "eye lost itself in inconceivable depths" of the ship, a symbolic "bottomless black pit," but the nautical instrument that *Hollow Sea* disparages ultimately orients him metaphysically on this voyage: "The road would be long . . . but this road my mind's eye could see on a chart, professionally, with all its complications and difficulties, yet simple enough in a way. One is a seaman or one is not."[17] And in *The Nigger of the Narcissus* the ship's stasis is attributed to the weather but also, narrative logic implies, to the work stoppages of the Afro-Caribbean James Wait and the labor agitator Donkin. The perspectives of the author whose aim is "to make you *see*" and provide a "glimpse of truth," proved problematic to interwar sea-fiction writers.[18] As Hanley's contemporary George Garrett argued: "Seeing . . . does not always exclude personal preferences. And personal preferences must not pass as whole truth. Conrad had his. . . . He wrote as a conservative-minded ship's officer"[19] who turned working-class mariners such as Donkin into "scapegoats." Garrett concludes that one day "Donkin might write the story of the sea. Let's hope it will be to a better world in which ship-owners can still send out heavily insured coffin ships and their helpless crews."[20] Hanley departs from Conrad by making us see the hydrosphere through the eyes of mariners across all classes and positions in the hydrarchy aboard a "coffin" or death ship while expanding this term's meaning beyond that of a declining shipping industry's dilapidated merchant vessels to refer to merchant vessels commandeered by the state.

Hollow Sea is a modernist sea narrative as Cesare Casarino defines it: a work in which "life aboard the ship becomes the central telos of the narrative and is revealed in all of its explosive economies of power," including "its disciplinary mechanisms . . . hierarchical subdivisions and distributions of space, the whole multiform dialectic of capital and labor."[21] In fact, Hanley formally attributes the clearest sight, the sharpest and fullest impressions, to those whose social status is lower on the hydrarchy but who occupy the space on the ship that is high above all others.

Hollow Sea intervenes in the top-down, panoptic power wielded by the "ethereal voice" by countering the state's war optics with the privileging of a key figure in maritime history and fiction: the lookout man. Of the latter character, Rochdale, the text relates: "He was powerful. The power lay behind the eyes. His strong hands and body were useless now. The eyes and the power and intelligence behind them were everything" (34). Hester Blum has theorized the literal and figural meanings of this "sea-eye," foregrounding its importance to revising the history of the age of sail through the "analytical perspective of the common laboring sailor."[22] Looking to views "from the masthead," she finds a materialist epistemology "by which the practices of mechanical labor . . . enable moments of reflection and speculation . . . yielding a generalized form of nautical experience."[23] Correcting the limited vision of the Admiralty and the upper officer with the wide, sharp, scopic control of the working classes, *Hollow Sea* makes this figure its "sea-eye." Ultimately it is the lookout man, not the captain and not the Admiralty, who discloses to readers what none of these authorities is able to witness in real time:

> Rochdale saw it all. He was high in the air, safe in his fastness. He saw everything. He saw the transport free of its burden now, turning slowly away, saw the grey ships go in nearer and nearer, the circle of fire spread. He saw boats descend and ascend, heard shouts and screams, the hissing of rope, wood hurtle in the air, and fall in great splinters when they were engulfed, as those bodies were engulfed. He had seen them rise, sink, rise again, the waters threshed, saw the tattered ribbon that weaved its way towards the bloody beach, and that was human and the name was written upon waters, courage, madness. (218–19)

By situating the working-class merchant mariner as the witness to war trauma, *Hollow Sea* departs from and adds to modernist representations that focalize through soldiers returned from the front, having experienced trench warfare. By privileging this view from the masthead, Hanley's novel not only deepens our understandings of the relationships of modernism to World War I but challenges dominant optics of both maritime literature and the state.

NICOLE RIZZUTO is associate professor of English at Georgetown University. She is author of *Insurgent Testimonies: Witnessing Traumatic History in Modern and Anglophone Literature* (2016), which was short-listed for the Modernist Studies Association Prize for a First Book. Her new book examines stasis and intercepted mobility as structuring principles of Anglophone literature and art from the interwar period through today. She has published essays examining the topic of stalled movement at sea in the writings of Virginia Woolf and Richard Hughes. These appear in *Modernist Cultures* and the print plus platform of *Modernism/Modernity*.

Notes

1 See Peter Linebaugh and Marcus Rediker's analyses of hydrarchy, transnational community, and democracy at sea in the seventeenth and eighteenth centuries (*Many-Headed Hydra*); Cesare Casarino's reconstellation of the works of Marx, Melville, and Conrad within the hydrosphere (*Modernity at Sea*); Ian Baucom's wide-ranging study of global capitalist modernity anchored to a single, traumatic event at sea, the Zong Massacre of 1781 (*Specters of the Atlantic*); and Sowande' Mustakeem's reconstruction of the Middle Passage from a multitude of maritime texts and documents (*Slavery at Sea*). Postcolonial critics have long considered how far-flung waters figure in literatures of decolonization, migration, and neocoloniality, among them Elizabeth DeLoughrey and Brian Bernards.

2 Jonathan Raban's introduction and selections in *The Oxford Book of the Sea*, for example, tacks from Renaissance through high modernist works only to move immediately to the postwar fiction of Monserrat. Margaret Cohen's encyclopedic study *The Novel and the Sea* spans the seventeenth century through the first decade of the twentieth, then looks ahead in an epilogue to postwar and twenty-first-century works.

3 Fussel, *Abroad!*

4 Schmitt, *The* Nomos *of the Earth*, 235.
5 Deer, *Culture in Camouflage.*
6 Saint-Amour, "Modernist Reconnaissance," 349–50.
7 Cohen, *The Novel and the Sea*, 180.
8 Other studies that treat *Heart of Darkness* as a form of maritime fiction broadly conceived include Duffy, *Speed Handbook*, 86–94, which reads it as a story of slowness and intercepted progress emblematized in part by the stalled and failing ship, and Casarino, *Modernity at Sea*, which links it to *The Nigger of the Narcissus* by way of the idea of the heterotopia of the ship.
9 Hanley, *Hollow Sea*, 21 (hereafter cited by page number).
10 On the relationships between colonial spaces and the sea as "free spaces" beyond amity lines, see Schmitt, *The* Nomos *of the Earth*, 94.
11 Conrad, *Heart of Darkness*, 10.
12 Friedman, *Network-Centric Warfare*, x.
13 Friedman, *Network-Centric Warfare*, 3.
14 Sondhaus, *Great War at Sea*, 180.
15 Conrad, *Typhoon*, 29, 30.
16 Conrad, *Shadow-Line*, 51.
17 Conrad, *Shadow-Line*, 88, 41.
18 Conrad, *Nigger of the Narcissus*, ii.
19 Garrett, *Collected George Garrett*, 245.
20 Garrett, *Collected George Garrett*, 245.
21 Casarino, *Modernity at Sea*, 9.
22 Blum, *View from the Masthead*, 133.
23 Blum, *View from the Masthead*, 109.

Works Cited

Baucom, Ian. *Specters of the Atlantic: Finance Capital, Slavery, and the Philosophy of History.* Durham, NC: Duke University Press, 2005.

Bernards, Brian. *Writing the South Seas: Imagining the Nanyang in Chinese and Southeast Asian Postcolonial Literature.* Seattle: University of Washington Press, 2016.

Blum, Hester. *The View from the Masthead: Maritime Imagination and Antebellum American Sea Narratives.* Chapel Hill: University of North Carolina Press, 2008.

Casarino, Cesare. *Modernity at Sea: Melville, Marx, Conrad in Crisis.* Minneapolis: University of Minnesota Press, 2002.

Cohen, Margaret. *The Novel and the Sea.* Princeton, NJ: Princeton University Press, 2010.

Conrad, Joseph. *Heart of Darkness*, edited by Robert Kimbrough. New York: Norton, 2006.

Conrad, Joseph. *The Nigger of the Narcissus*, edited by Robert Kimbrough. New York: Norton, 1979.

Conrad, Joseph. *The Shadow-Line: A Confession*, edited by J. H. Stape and Allan H. Simmons. Cambridge: Cambridge University Press, 2013.

Conrad, Joseph. *Typhoon and Other Tales*, edited by Cedric Watts. Oxford: Oxford University Press, 2008.

Deer, Patrick. *Culture in Camouflage: War, Empire, and Modern British Literature.* Oxford: Oxford University Press, 2009.

DeLoughrey, Elizabeth M. *Routes and Roots: Navigating Caribbean and Pacific Island Literatures.* Honolulu: University of Hawai'i Press, 2007.

Duffy, Enda. *The Speed Handbook: Velocity, Pleasure, Modernism.* Durham, NC: Duke University Press, 2009.

Friedman, Norman. *Network-Centric Warfare: How Navies Learned to Fight Smarter through the World Wars.* Annapolis, MD: Naval Institute Press, 2009.

Fussel, Paul. *Abroad! Literary Traveling between the Wars.* Oxford: Oxford University Press, 1982.

Garrett, George. *The Collected George Garrett*, edited by Michael Murphy. Nottingham: Trent, 1999.

Hanley, James. *Hollow Sea.* London: Kimber, 1976.

Linebaugh, Peter, and Marcus Rediker. *The Many-Headed Hydra: The Hidden History of the Revolutionary Atlantic.* London: Verso, 2012.

Mustakeem, Sowande'. *Slavery at Sea: Terror, Sex, and Sickness in the Middle Passage.* Urbana: University of Illinois Press, 2016.

Raban, Jonathan. *The Oxford Book of the Sea.* New York: Oxford University Press, 1992.

Saint-Amour, Paul K. "Modernist Reconnaissance." *Modernism/Modernity* 10, no. 2 (2003): 349–80.

Schmitt, Carl. *The* Nomos *of the Earth: The International Law of the* Jus Publicum Europaeum. New York: Telos, 2006.

Sondhaus, Lawrence. *The Great War at Sea: A Naval History of the First World War.* Cambridge: Cambridge University Press, 2014.

"We Cannot Think of a Time That Is Oceanless"

Oceanic Histories in "The Dry Salvages"

MAXWELL UPHAUS

Abstract This essay proposes, through an analysis of T. S. Eliot's "The Dry Salvages," a model for the study of the sea in modernism based on British modernism's relationship with the contemporaneous idea that the sea was the essence of British history. "The Dry Salvages" rejects the view of the sea as embodying British history entertained elsewhere in Eliot's poetry of this period, but it continues to envision Britain as inherently caught up in the new vision it elaborates of oceanic history as an ongoing calamitous process that is nevertheless essential to historical thinking as such. The poem develops this new vision by both suppressing and drawing on Black Atlantic historical experience. As it undertakes this oceanic historical reconceptualization, "The Dry Salvages" demonstrates how, in British modernism, thinking about the sea and thinking about history are intrinsically interconnected.
Keywords modernism, oceanic histories, maritime imperialism, Britishness, the Black Atlantic

"The Dry Salvages," the third of T. S. Eliot's *Four Quartets,* not only rewards a hydro-critical reading; it also figures prominently in the development of hydro-criticism itself. Eliot's poem provides epigraphs or quotations for a series of major contributions to the study of the sea and the transnational connections it facilitates. These studies form a chain of direct or indirect influence, stretching from C. L. R. James in 1962 to Peter Linebaugh in 1982 to Elizabeth DeLoughrey in 2010.[1] None of these writers engages substantively with Eliot's poem, but all of them present it as informing or epitomizing their own ideas. So what is it that makes "The Dry Salvages"—a poem by a white royalist Anglo-Catholic, and part of a sequence that seemingly embraces a conservative, insular version of English national identity—so useful, even formative, for a succession of left-wing, anti- or postcolonial scholars of the Atlantic world, representing one of the main genealogies of oceanic studies?

57:1, April 2019 DOI 10.1215/00138282-7309743

There are several answers to this question, some more obvious than others. By reflecting on the sea that connects Britain and America, "The Dry Salvages" showcases the sea's role as a means of transcending national frames: a hallmark of oceanic studies since at least Paul Gilroy's call for an ocean-focused theoretical perspective that would be "less intimidated by and respectful of the boundaries and integrity of modern nation states."[2] The poem further anticipates present-day hydro-critical concerns by portraying the ocean as an immense nonhuman agent whose spatiotemporal scale vastly outstrips that of human beings and demands a fundamental reconsideration of humanity's place in the world.[3] "The Dry Salvages," indeed, stresses the ocean's nonhuman agency and all-pervasive presence to the extent of blurring the very boundary between sea and land.[4] Most important for my argument, Eliot's poem foregrounds the ocean's multifaceted historical significance: the way in which it can be seen both to animate the histories of nations, empires, and civilizations and to assert their inconsequence in a geohistorical perspective. "The Dry Salvages," that is, envisions the sea much as a range of subsequent oceanic writing and scholarship does: as simultaneously making, memorializing, and eclipsing human histories, a status encapsulated by Derek Walcott's famously ambiguous avowal "The sea is History."[5]

However, "The Dry Salvages" approaches these concerns through the prism of an ideology that pervaded British life during the half century preceding the poem's appearance: the idea that the sea was essential to British history, the element that made that history possible and in which it was enshrined.[6] We cannot fully understand the poem's deployment of the ocean, I argue, without first understanding its relationship to this belief that the sea activated, propelled, and embodied British history. "The Dry Salvages" incorporates this ideology in a complex, dynamic way, perpetuating it in some respects while counteracting it in others. The poem alludes to the idea that the sea forms an animating pathway for British expansion; moreover, it suggests that looking to the sea allows one to see beyond differences in the phase and mode of expansion to discern the fundamental unity of such expansionary history from the mid-seventeenth to the mid-twentieth centuries. This engagement with maritime history, though, is inflected by the wartime context of the poem's composition and publication in 1940–41, when Germany's U-boat campaign caused a crisis on the Atlantic that called into question whether the ocean could continue to be envisioned as a highway for and repository of British expansionary history. "The Dry Salvages" stages a transition, under the pressure of this crisis, from an idea of the sea as driving and enshrining British history to a view of the ocean as an entity circumscribing and exceeding humanity, in whose tumultuous nonhuman history we are engulfed. This conceptual shift that the poem enacts impels, in turn, a retreat to an insular English identity. Yet the poem also portrays this England as still situated within an oceanic history that permeates and even constitutes that insular identity.[7]

Eliot's poem thus unfolds a complex view of the ocean as integral to British history and as a force that exceeds and engulfs it. As it does so, the poem develops an additionally complex relationship with another aspect of the sea's historicity: its capacity to "memorialize histories of violence," especially the "wasted lives in the

Middle Passage."[8] "The Dry Salvages" problematically naturalizes such violent Black Atlantic experience; simultaneously, however, this experience informs the new view of oceanic history—as an ongoing calamitous process that is nevertheless essential to historical thinking as such—that the poem elaborates.[9] It is above all for this reason, I suggest, that the poem has proved so influential for generations of Black Atlantic and oceanic studies scholarship. As well as thus clarifying the poem's importance for hydro-criticism, tracing the interacting oceanic histories in "The Dry Salvages" demonstrates how, in the poem, oceanic representation and historical thought are intrinsically interconnected—or, as the poem puts it, "We cannot think of a time that is oceanless."[10]

"The Dry Salvages" reflects a moment when, for the second time in Eliot's life, the ocean linking him to his American origins had become a battlefield. The beginning of 1941 saw Germany intensify its U-boat offensive against British shipping, threatening the maritime lifelines that held Britain's empire together and enabled it to stand alone against the Axis. By the end of February, Britain's imperiled sea lanes and overstrained maritime infrastructure were near collapse.[11] This crisis clearly affected Eliot—all the more so given the impact that the first round of U-boat warfare during World War I had had on him a quarter century before.[12] His concern found expression in "The Dry Salvages," first published on February 27. The poem's maritime descriptions repeatedly evoke what would shortly be termed "the Battle of the Atlantic."[13] "The Dry Salvages" directs attention to the perils faced by convoys and their escorts—"those concerned with every lawful traffic / And those who conduct them" (197)—on "an ocean . . . littered with wastage" (193), including the "wastage" of sunken merchantmen and warships, and it summons its readers to renewed effort in response: "Not fare well, / But fare forward, voyagers" (197).[14]

These evocations of the Battle of the Atlantic were not the first time, even during this new conflict, that Eliot's poetry had reflected on the war at sea or urged fresh commitment in the face of its losses. In June 1940 Eliot wrote a poem, "Defence of the Islands," to accompany a planned exhibition of British war photographs in New York.[15] In its list of Britain's defenders, the poem gives pride of place to

> those appointed to the grey
> ships—battleship, merchantman, trawler—
> contributing their share to the ages' pavement
> of British bone on the sea floor.
>
> (213)

This depiction of British seafaring perpetuates a view of the sea as constituting British history. The poem rehearses a central trope of this ideology as it had been articulated since at least the late nineteenth century: the notion that all the Britons who died at sea during the nation's centuries of seafaring had made the sea itself British, an embodiment of Britain's past. Eliot's "ages' pavement / of British bone" echoes, for example, Robert Louis Stevenson's 1878 account of "the pretension that the sea is English": "Even where it is looked upon by the guns and battlements of another nation we regard it as a kind of English cemetery, where the bones of our seafaring fathers take their rest until the last trumpet; for I suppose no other nation has lost as many ships, or sent as many brave fellows to the bottom."[16] Or as Rudyard Kipling writes of the sea in an 1896 poem Eliot included in his *Choice of Kipling's Verse*, published later in 1941: "There's never a wave of all her waves / But marks our English dead."[17]

In thus depicting it as a graveyard of British mariners, Stevenson and Kipling frame the sea as an extension of Britain—another foreign field that is forever England. More, they cast the sea as an inherently historical space. From these writers' perspective, by enshrining "the bones of our seafaring fathers," the sea bears enduring witness to the expansionary history those forefathers enacted and materially embodies what made Britain what it is. Similarly, the sea of "Defence of the Islands" embodies the "ages" of British maritime enterprise through its sediment of "British bone," thereby giving an enduring form to British history. For Eliot, as for the earlier writers, this materialization of maritime history legitimates Britain's empire: sunken British bone figuratively paves the way for the ongoing imperial voyaging of "the grey ships" while also making it easier to imagine a territorial continuity of Britishness from "the islands" to their possessions overseas that stretches across, or rather beneath, the ocean.

Where Stevenson's and Kipling's works reflect an expanding Britain, of course, Eliot's poem is overtly defensive—a distinction that highlights the difference between high Victorian imperialism and the empire's structure and situation in 1940. Yet Eliot's recycling of a trope from that earlier period in these altered circumstances underscores an idea that his poem itself suggests, which is that a maritime perspective reveals the underlying unity of British imperial history. The poem presents the expansion that has left the seafloor paved with British bone as an ongoing, accumulative process, to which "battleship, merchantman, [and] trawler" all "contribut[e] their share." That is, the kinds of activity and forms of power represented by the battleship (maritime military power), the merchantman (maritime commercial capitalism), and the trawler (the extraction of resources from the sea) all play a part in the oceanic extension of Britishness. And these kinds of seafaring did in fact interact historically as successive, complementary modes of spreading British influence by sea: England's growing merchant marine from the sixteenth century on fed the rise of British naval power and long overlapped with it, while that mercantile expansion itself followed on the voyages of the fishermen who first ventured out into the Atlantic.[18] Eliot's alignment of fishing boats, commercial vessels, and warships thus suggests a fundamental continuity through superficial difference: a consistent process of seaborne expansion to which each mode of seafaring, and the mode of imperial power projection associated with it, contributes its share.

Like Stevenson's and Kipling's, however, Eliot's depiction of the sea in these terms—as an embodiment of British maritime-imperial history that reveals that history's essential continuity—requires overlooking the sea's status as a repository of the violence of this expansionary history: in particular, the millions of enslaved Africans "sent . . . to the bottom" in the Middle Passage. In awarding the sea to Britain based on the quantity of British dead it contains, "Defence of the Islands" joins its predecessors in marginalizing such maritime atrocities and suppressing the claim of the people who suffered them, and their descendants and inheritors, to the historical space of the sea. Yet like its predecessors, Eliot's poem also displays a desire to portray Britons as not just the agents but also the victims of their own seaborne expansion, as well as a sense that victimization confers a stronger claim to the sea than victory does. In other words, the poem needs not only to suppress but also to supplant the dead of the Middle Passage. This need bespeaks an underlying recognition of the enormity of Atlantic slavery and its centrality to Britain's oceanic history. Even Kipling cannot reflect on maritime Britishness without giving a stanza, in another 1896 poem Eliot included in his *Choice of Kipling's Verse*, to "the souls of the slaves that men threw overboard." Kipling ventriloquizes these Middle Passage dead only to turn their destruction to a redemptive end—"Thy [God's] arm was strong to save, / And it touched us on the wave"—and his appropriation of their voices replicates the violence perpetrated against them.[19] At the same time, his example testifies to the haunting presence of this dimension of oceanic British history even at the high tide of maritime empire, and to the necessity of addressing this presence in some way, however exculpatory. We will see a similar imperative at work in "The Dry Salvages," in a manner that reflects the altered conditions of maritime empire in 1941.

Underlying "The Dry Salvages" is a vision of the sea similar to that of "Defence of the Islands": a space that preserves the legacy of "seafaring fathers" and the "ages" of British expansion, makes visible the essential continuity of that expansion, and thereby embodies what has made Britain what it is. The poem moves from the English settings of the previous two quartets to the sites of Eliot's American upbringing: St. Louis, remembered in the opening stanza, and especially the Massachusetts coast, one feature of which gives the poem its title. This personal return to America also reenacts the seventeenth-century transatlantic voyage of Eliot's ancestors. The previous quartet, "East Coker," departs from the village of that name, where the Eliot family originated, to set forth on "the vast waters" (190) —like Eliot's forebear Andrew Eliott sailing for Massachusetts in 1669. "The Dry Salvages" then follows this ancestral sea journey to its American destination.[20] Familial history also merges with national history, insofar as the migration of Eliot's "seafaring fathers" formed part of England's broader transformation from an agrarian island kingdom to a mercantile seaborne empire. By reenacting this familial and national transition of three centuries before, "The Dry Salvages" embeds a view of the sea as an element that animates and embodies history.

Nor is the seventeenth-century takeoff of English seaborne colonization and trade the only stage of British expansion the poem conjures. In part 2 the poem glances at the dark side of that seventeenth-century expansionary surge in a line recalling the flooded Mississippi of Eliot's St. Louis childhood: "the river with its cargo of dead negroes" (195). Although it stems from a different place and time, this specter of a waterborne "cargo of dead negroes," read in the light of the "ocean . . . littered with wastage" (193) that the poem has just described, calls to mind the Middle Passage—the other major vector of transatlantic migration at the time

Andrew Eliott made his voyage, and the central leg of the Atlantic triangular trade that contributed immensely to Britain's growing commercial, financial, and imperial power between the seventeenth and the nineteenth centuries. Part 3 of the poem then develops an extended reference to the Bhagavad Gita that brings into view another zone of British expansion: India. Indeed, when it represents the essence of this Hindu philosophical text as "a voice descanting" on a "drumming liner" in midocean—"here between the hither and the farther shore" (196)—the poem undertakes a seaborne passage to India in a manner that evokes the maritime basis of British rule there from the eighteenth century to Eliot's time. The poem's oceanic perspective aligns these phases of British imperial history as distinct but related forms of maritime expansion. Looking to the sea, "The Dry Salvages" proposes, allows one to see an underlying continuity among the colonization of America, the triangular trade, imperialism in India, and the Battle of the Atlantic—to understand how these different forms of imperial activity, like the battleships, merchantmen, and trawlers of "Defence of the Islands," form complementary aspects of a common oceanic history.

Yet while this conception of the sea as propelling, embodying, and revealing the unity of imperial history underlies "The Dry Salvages," the poem also severely qualifies such a conception almost immediately. Primed by the reenactment, at the end of "East Coker," of Britain's historical turn to the sea, the reader begins "The Dry Salvages" expecting to see the sea portrayed as a highway to empire. However, the poem's first reference to the sea upsets this expectation: "The river is within us, the sea is all about us" (191). In this line the river is internal, implicitly humanized, but the sea is external—a circumscribing element set apart from "us." After seemingly leaving the island with Eliot's seafaring ancestors, we find ourselves right back in insularity. The ensuing lines augment this impression of the sea as other. The first signs the sea leaves, before it is seen to bear any traces of humanity, are "its hints of earlier and other creation: / The starfish, the horseshoe crab, the whale's backbone" (191). These lines arouse a disquieting sense of the sea as the product of a different creation than the one that led to humanity—as, in effect, an alien universe. Instead of preserving "British bone," the sea here washes up whalebone; instead of enshrining maritime-imperial history, it attests to a deeper, nonhuman history.

The poem then equates the ocean and nonhuman time openly—not just in its content but also in its meter and form. By the time it first shifts its focus to the sea, "The Dry Salvages" has adopted a relatively regular, four-stress line: "The ríver is withín us, the séa is all abóut us." As Eliot emphasizes the sea's abyssal timescale, though, this metrical regularity drifts into a tidal dilation and contraction:

> And under the oppression of the silent fog
> The tolling bell
> Measures time not our time, rung by the unhurried
> Ground swell, a time
> Older than the time of chronometers.
>
> (192)

This attempt to "measure" an inhuman oceanic temporality with an irregular meter comes to a head in the jarringly uneven concluding lines of part 1, which move from thirteen syllables to one and then to two. This lapse out of a regular metrical pattern further emphasizes oceanic resistance to human temporal schemes. Through this thematic and formal emphasis on the discordance between our time and the ocean's, "The Dry Salvages" adopts a dramatically different view of the sea's relationship with history from that of "Defence of the Islands." "Our time" is precisely what the sea "measures" in that earlier poem, insofar as its "pavement" of British bone records the ages of Britain's existence as a seafaring nation. Such a vision of the sea as embodying human historical time vanishes in part 1 of "The Dry Salvages." The poem instead associates the oceanic with the geohistorical, the timescale of the ocean with the timescale of the planet itself.

Having thus departed from a view of the sea as embodying British history, "The Dry Salvages" recontextualizes the maritime crisis of its moment within the ocean's "time not our time." The poem naturalizes "the drifting wreckage" (193) of the Battle of the Atlantic as part of an immemorial condition, the state of inherent, destabilizing flux that has always been the ocean's real history. Rather than a consequence of specific human acts in a specific political and historical situation, an ocean littered with wastage comes to seem beyond human cause or control—something permanent and ineluctable. In this, Eliot's poem reflects the fact that the kind of crisis it responds to had already happened once in his own and Britain's experience, and that loss and destruction on the waves the nation once professed to rule were beginning to be taken for granted.[21] "The Dry Salvages" enshrines this view by subsuming the current crisis on the Atlantic, and the centuries of British maritime-imperial history that preceded it, into a single continuous catastrophe: "the drift of the sea and the drifting wreckage," to which, the poem affirms, "there is no end" (194). The solidity of "the ages' pavement / of British bone" dissolves into ongoing drift, while the progressive sedimentation of bones and ages becomes an unbounded accumulation of wreckage and time: "There is no end, but addition" (193).[22] If the sea is history in "The Dry Salvages," in short, it is a nonhuman history that swamps "the assurance / Of recorded history" (195).

In the second half of part 2 the poem distills this antiprogressive conception of history in its memory of "the river with its cargo of dead negroes." This image highlights both the implicit effects and the hidden sources of Eliot's reconceptualization of oceanic history. As argued above, the image evokes the Middle Passage, but it also marginalizes this experience and naturalizes it in the same way as the poem does the Battle of the Atlantic and other oceanic disasters. This suggestion of the Atlantic slave trade is even more oblique than Eliot's other historical allusions: rather than pointing directly to Middle Passage atrocities, the poem moves "its cargo of dead negroes" from the Atlantic to the Mississippi and up in time to Eliot's own childhood. As it does so, the poem turns the literal reality of Middle Passage cargoes into a metaphor, and it presents the metaphorical shipment and actual death of the "negroes" it envisions as natural phenomena, due not to human action but to the natural force of the river. Although the vulnerability to flood of St. Louis's black population had thoroughly human and historical causes, "The Dry Salvages" simply takes the resulting deaths for granted as an inherent aspect of existence. The poem by extension also naturalizes the dead of the Middle Passage that Eliot's image dimly recalls. Like "Defence of the Islands," "The Dry Salvages" thus minimizes the violence of British maritime expansion, though in a different way—not by omitting it from a propagandistic picture of the sea enshrining British history but by subsuming it in a poetic vision of an inherently destructive oceanic history.

Yet this naturalization of such waterborne racial violence also attests to the poem's need to do something with this legacy—its inability to keep it completely beneath the surface, any more than Kipling could survey British maritime-imperial history without trying to redeem the murder of "the slaves that men threw overboard." "The Dry Salvages" likewise cannot encapsulate three centuries of British maritime expansion without registering in some form that history's constitutive concomitant: Atlantic slavery. As Linebaugh puts it in an article that uses as an epigraph Eliot's passage envisioning the "cargo of dead negroes," "Whatever high points stood out in the sun of European imperialism, they always cast an African shadow."[23] In "The Dry Salvages" the "sun" of the British seaborne expansion from the seventeenth century on that underlies the poem casts as its "shadow" this specter of the aquatic destruction of black bodies. Moreover, by embedding its evocation of such destruction in the wake of its portrayal of an ongoing oceanic history of "wastage," the poem suggests that the Middle Passage underlies its reconceptualization of oceanic history. Amid the maritime-imperial crisis of early 1941, that is, as Britain's ability to rule the waves comes into doubt, the violence that formerly enabled it to do so gains renewed salience and contributes to Eliot's bleaker outlook. The Battle of the Atlantic brings the Atlantic slave trade into view, however dimly, and while the poem minimizes the experience of slavery it recalls, this recollection also fuels the poem's altered view of what an oceanic history is and means. "The Dry Salvages" attempts to naturalize these two poles of British maritime history, the Battle of the Atlantic and Atlantic slavery, but by thus reenvisioning oceanic history as an ongoing, catastrophic process that leaves an ocean strewn with wastage, it also reflects, connects, and universalizes these historical episodes. Each draws out aspects of the other, and the poem's attempt to explain them away also has the effect of keeping them in view.

Eliot's reimagining of the sea's historical role in these catastrophic terms would seem to restore a familiar land/sea binary, in which the sea is insecure, unknowable, and alien while the land is stable, comprehensible, and humane. Such a binary would seem to reorient Britishness from the sea to the land, paralleling the turn from imperial expansion to insular nationality that has been ascribed to *Four Quartets*.[24] In one sense, this is exactly what "The Dry Salvages" accomplishes. The poem's engagement with the unbounded flux of oceanic time leads to a retreat, in its final lines, to the terrestrial stability of "significant soil" (199). And in the same way that "East Coker's" concluding turn to "the vast waters" set up "The Dry Salvages," this closing reversion to "significant soil" sets up the recourse to insular Englishness in the final quartet, "Little Gidding," in which a terrestrial English locale takes the place of the ocean as the embodiment of English history: "History is now and England" (208).

However, "The Dry Salvages" also dissolves the land/sea binary. The poem suggests that the land cannot be securely distinguished from the sea that surrounds and sculpts it. The poem's first lines describing the sea affirm the sea's power to both mold and erode, which permeates and, it seems, even constitutes dry land:

> The sea is the land's edge also, the granite
> Into which it reaches, the beaches where it tosses
> Its hints of earlier and other creation.
>
> (191)

Here the sea does not just penetrate but *is* dry land, even "granite"—an unsettling liquefaction of terra firma that the enjambment on *granite* accentuates. Similarly, the internal rhyme of "reaches" and "beaches" evokes interpenetration and transformation: the beaches both emerging from and slipping back into the sea that reaches. The line between sea and land is here as thin as a single initial consonant. Shortly afterward a half line—emphasized by being offset from the rest of its line—reiterates the extent of the sea's reach: "The salt is on the briar rose" (192). There is no escaping the sea; it leaves its mark everywhere, even on the "rose" that will become "Little Gidding's" ultimate apocalyptic emblem. The "significant soil" on which "The Dry Salvages" ends up taking refuge comes to seem only the impermanent solidification of an essential oceanic reality.

In blurring the boundary between sea and land, "The Dry Salvages" also undermines an insular version of English nationality. If the solid ground of the island cannot be separated from the ocean, then, by extension, the island nation cannot be separated from its oceanic history. "The Dry Salvages" thus continues to suggest that the sea constitutes British identity, but it reconfigures this constitutive relationship between Britain and the sea. Instead of being upheld by an oceanic history, Britain in this new conception is subordinated to it as the maritime-historical currents that once conveyed its expansion now flow back to and reshape it.

"The Dry Salvages" suggests as much in a line in its final section that gestures again at the poem's wartime context: its composition during a time "when there is distress of nations and perplexity / Whether on the shores of Asia, or in the Edgware Road" (198). On the surface, the line stresses the distance between Asia and London: *only* common "distress" and "perplexity" unite these otherwise far-flung places. However, the alliteration of *A*sia and *Ed*gware aurally connects the two. Asia is linked with and embedded in the Edgware Road in Eliot's poetic line, just as it increasingly was in life. (The poem's incorporation of the Bhagavad Gita in part 3 enacts a similar interconnection.) The reference to "shores" and the *Edg* in *Edgware* also recalls the poem's earlier description of the ocean's penetration into dry land: "The sea is the land's *edge* also, . . . / . . . the beaches where it tosses / Its hints of earlier and other creation." In the same way, the poem intimates, Britain is now being pervaded and remade by the people and influences that its maritime-imperial history has brought to it. Britain's oceanic highway to empire has reversed direction—or rather, oceanic history was never a highway at all. Instead, it was the ongoing chaotic process the poem has described, one that Britain formerly could dominate but that is now fundamentally altering Britain itself. In this manner, "The Dry Salvages" continues to position Britain within its more volatile version of oceanic history: instead of the sea being an extension of Britain, Britain becomes an emergence from the sea.

Hence, even as it departs from an imperialistic view of the sea's historical role, "The Dry Salvages" affirms the historical importance of the ocean and the oceanic nature of history. The poem encapsulates its new vision of the sea and history in the pair of lines that DeLoughrey's 2010 hydro-critical article takes as its epigraph: "We cannot think of a time that is oceanless / Or of an ocean not littered with wastage" (193). These lines recast the seafloor paved with British bone as a wastage-strewn ocean, and Britain's maritime-imperial "ages" as a geohistorical "time." While this recasting may make oceanic time both more difficult to conceptualize and more disquieting to contemplate, however, it also makes the connection of the ocean to a time that we can think of even more fundamental. As the Battle of the Atlantic rages and the end of empire looms, the sea can no longer easily be envisioned as a highway for or archive of imperial expansion, but its indispensability for historical thinking only deepens: in other words, the sea remains history. And in working out this new sense of the sea as history, "The Dry Salvages" both draws on and minimizes specific oceanic histories of displacement and violence—in particular, Atlantic slavery. If it pushes them below the surface, the poem's oceanic historical reconceptualization also reflects and points to these histories. Eliot's poem incorporates the Black Atlantic history it subordinates—sufficiently, at least, for subsequent scholars to use the poem to epitomize such diasporic maritime historical experience. Rereading "The Dry Salvages" in these terms thus not only elucidates the poem's use of the sea to think through British history in a manner that questions both maritime imperialism and insular nationalism; it also clarifies how and why the poem has helped animate a trajectory of writing and thinking about the sea that continues today.

MAXWELL UPHAUS is assistant professor of English specializing in British modernism at the University of Toronto. His research explores how modernist writers from Britain and its empire envisioned and represented history by drawing on different ideas of the ocean's historical status and role. His recent work has been published in *Modernist Cultures* and *Soundings: An Interdisciplinary Journal.*

Acknowledgments

I would like to thank Peter Luebke, Audrey Walton, and the anonymous reviewers for their comments and suggestions regarding prior versions of this article. The article draws on a presentation I gave for the panel "Eliot and Ecocriticism" at the 2018 Modern Language Association convention in New York; I am grateful to my fellow participants on the panel, especially its organizer, Julia Daniel.

Notes

1 James, *Black Jacobins*, 402; Linebaugh, "All the Atlantic Mountains Shook," 87; DeLoughrey, "Heavy Waters," 703. The same passage from "The Dry Salvages" that James quotes in *The Black Jacobins* also provides the title and epigraph of the second volume of his selected

writings (*Spheres of Existence*). For a more recent use of a passage from Eliot's poem to exemplify a new approach to the study of the sea in the humanities, see Miller, "Introduction," 19.

2 Gilroy, *Black Atlantic*, 4.

3 For a recent account of the ocean in these terms that draws on Timothy Morton's characterization of such massive nonhuman beings as "hyperobjects," see Mentz, *Shipwreck Modernity*.

4 For an oceanic-studies call to rethink the land/sea binary, see Steinberg, "Of Other Seas," 163–64.

5 Walcott, "The Sea Is History," in *Collected Poems*, 364. Eliot's importance for Walcott's literary development has been frequently noted, including by Walcott himself: see, for example, *Conversations with Derek Walcott*, 123, in which Walcott connects his excitement about discovering Eliot to his adolescent sense of belonging to "the heritage of the British Empire." See also Pollard, *New World Modernisms*. For representative scholarly studies of the ocean's historicity, see DeLoughrey, "Heavy Waters"; and Baucom, *Specters of the Atlantic*.

6 On the political and cultural importance of the sea and sea power in late nineteenth- and early twentieth-century Britain, see Behrman, *Victorian Myths of the Sea*; and Rüger, *Great Naval Game*. For recent work putting modernist depictions of the sea in dialogue with contemporaneous British maritime ideology, see Rizzuto, "Maritime Modernism"; and Uphaus, "Hurry Up and Wait."

7 My argument thus seeks to intervene in the long-standing debate over how *Four Quartets* envisions English national identity. For important recent entries in this debate, see Esty, *Shrinking Island*, 108–62; and Abravanel, *Americanizing Britain*, 131–56.

8 DeLoughrey, "Heavy Waters," 704, 708.

9 In this regard, "The Dry Salvages" exemplifies how, as Laura Doyle and Laura Winkiel put it, "canonical white Anglo modernism is . . . haunted by ghosts," including "the repressed ghosts" of "Atlantic modernity" ("Introduction," 3).

10 Eliot, *Collected Poems*, 193. Subsequent references to Eliot's poetry are to this edition, cited parenthetically by page number.

11 Terraine, *Business in Great Waters*, 308.

12 On the impact of World War I U-boat warfare on Eliot and his family, see the commentary on "Mr. Apollinax" in *The Poems of T. S. Eliot*, 1:434.

13 Winston Churchill officially proclaimed the Battle of the Atlantic in a directive issued just a week after Eliot's poem appeared (*Second World War*, 3:122–23).

14 On the connection between "The Dry Salvages" and the Battle of the Atlantic, see MacKay, *Modernism and World War II*, 86–88.

15 Marina MacKay has previously demonstrated this poem's relevance to "The Dry Salvages" (*Modernism and World War II*, 86–87). Eliot later insisted that he had never thought of this work as a "poem" at all; see the commentary in *The Poems of T. S. Eliot*, 1:1046–47. I use the spelling of the work's title as it appears in this volume.

16 Stevenson, "English Admirals," 36–37.

17 Kipling, *Choice of Kipling's Verse*, 91.

18 As the British maritime writer Frank Bullen pointed out in 1906, in the early days of English seafaring there was "no division of . . . vessels into warships on one side, and merchant ships or trading vessels exclusively on the other" (*Our Heritage the Sea*, 234). The influential naval historical theorist Alfred Thayer Mahan argued that Britain's large merchant marine continued to contribute to its naval power by imparting "staying power, or reserve force" (*Influence of Sea Power upon History*, 45). For an argument that fishing expeditions paved the way for early modern maritime exploration and expansion, see Fagan, *Fish on Friday*; for a similar argument from early twentieth-century Britain, see Williamson, *Voyages of the Cabots*. On a more immediate level, the alignment of battleships, merchantmen, and trawlers in "Defence of the Islands" alludes to the participation of British vessels of all types, military and civilian, in the Dunkirk evacuation, which took place shortly before the poem was written.

19 Kipling, *Choice of Kipling's Verse*, 81.

20 For a fuller argument that *Four Quartets* recapitulates the English colonization of America, see Abravanel, *Americanizing Britain*, 131–56.

21 According to the maritime cultural historian Duncan Redford, while in World War I unrestricted submarine warfare had "traumatised the British," the repetition of this threat in World War II "provoked a hostile, if muted, response" (*Submarine*, 127, 130). In other words, the cause of the Atlantic's being "littered with wastage" in early 1941 was no longer unprecedented and was, instead, starting to be accepted as a predictable feature of modern war—an important basis for the further naturalization that "The Dry Salvages" accomplishes.

22 In this respect, "The Dry Salvages" notably anticipates Ian Baucom's formulation, similarly informed by Atlantic historical experience, of a philosophy of history in which "time does not pass, . . . it accumulates" (*Specters of the Atlantic*, 34). But where Baucom's Atlantic temporality remains resolutely historical, Eliot's poem seeks to dissolve human historical specificity into its nonhuman oceanic *longue durée*.

23 Linebaugh, "All the Atlantic Mountains Shook," 109.

24 See, e.g., Esty, *Shrinking Island*, 108–62.

Works Cited

Abravanel, Genevieve. *Americanizing Britain: The Rise of Modernism in the Age of the Entertainment Empire*. Oxford: Oxford University Press, 2012.

Baucom, Ian. *Specters of the Atlantic: Finance Capital, Slavery, and the Philosophy of History*. Durham, NC: Duke University Press, 2005.

Behrman, Cynthia Fansler. *Victorian Myths of the Sea*. Athens: Ohio University Press, 1977.

Bullen, Frank. *Our Heritage the Sea*. London: John Murray, 1906.

Churchill, Winston. *The Second World War*. 6 vols. Boston: Houghton Mifflin, 1948–53.

DeLoughrey, Elizabeth. "Heavy Waters: Waste and Atlantic Modernity." *PMLA* 125, no. 3 (2010): 703–12.

Doyle, Laura, and Laura Winkiel. "Introduction: The Global Horizons of Modernism." In *Geomodernisms: Race, Modernism, Modernity*, edited by Laura Doyle and Laura Winkiel, 1–14. Bloomington: Indiana University Press, 2005.

Eliot, T. S. *Collected Poems, 1909–1962*. Orlando, FL: Harcourt, 1991.

Eliot, T. S. *The Poems of T. S. Eliot*, edited by Christopher Ricks and Jim McCue. 2 vols. Baltimore, MD: Johns Hopkins University Press, 2015.

Esty, Jed. *A Shrinking Island: Modernism and National Culture in England*. Princeton, NJ: Princeton University Press, 2004.

Fagan, Brian. *Fish on Friday: Feasting, Fasting, and the Discovery of the New World*. New York: Basic, 2006.

Gilroy, Paul. *The Black Atlantic: Modernity and Double Consciousness*. Cambridge, MA: Harvard University Press, 1993.

James, C. L. R. *The Black Jacobins: Toussaint L'Ouverture and the San Domingo Revolution*. 2nd ed. New York: Vintage, 1989.

James, C. L. R. *Spheres of Existence: Selected Writings*. London: Allison and Busby, 1980.

Kipling, Rudyard. *A Choice of Kipling's Verse, Made by T. S. Eliot, with an Essay on Rudyard Kipling*, edited by T. S. Eliot. London: Faber and Faber, 1941.

Linebaugh, Peter. "All the Atlantic Mountains Shook." *Labour/Le travail*, no. 10 (1982): 87–121.

MacKay, Marina. *Modernism and World War II*. Cambridge: Cambridge University Press, 2007.

Mahan, Alfred Thayer. *The Influence of Sea Power upon History, 1660–1783*. 12th ed. Boston: Little, Brown, 1918.

Mentz, Steve. *Shipwreck Modernity: Ecologies of Globalization, 1550–1719*. Minneapolis: University of Minnesota Press, 2015.

Miller, Peter N. "Introduction: The Sea Is the Land's Edge Also." In *The Sea: Thalassography and Historiography*, edited by Peter N. Miller, 1–26. Ann Arbor: University of Michigan Press, 2013.

Pollard, Charles W. *New World Modernisms: T. S. Eliot, Derek Walcott, and Kamau Brathwaite*. Charlottesville: University of Virginia Press, 2004.

Redford, Duncan. *The Submarine: A Cultural History from the Great War to Nuclear Combat*. London: Tauris, 2010.

Rizzuto, Nicole. "Maritime Modernism: The Aqueous Form of Virginia Woolf's *The Waves*." *Modernist Cultures* 11, no. 2 (2016): 268–92.

Rüger, Jan. *The Great Naval Game: Britain and Germany in the Age of Empire*. Cambridge: Cambridge University Press, 2007.

Steinberg, Philip E. "Of Other Seas: Metaphors and Materialities in Maritime Regions." *Atlantic Studies* 10, no. 2 (2013): 156–69.

Stevenson, Robert Louis [R. L. S.]. "The English Admirals." *Cornhill Magazine*, July 1878, 36–43.

Terraine, John. *Business in Great Waters: The U-Boat Wars, 1916–1945*. London: Cooper, 1989.

Uphaus, Maxwell. "Hurry Up and Wait: *The Nigger of the 'Narcissus'* and the Maritime in Modernism." *Modernist Cultures* 12, no. 2 (2017): 173–97.

Walcott, Derek. *Collected Poems, 1948–1984*. New York: Farrar, Straus and Giroux, 1986.

Walcott, Derek. *Conversations with Derek Walcott*, edited by William Baer. Jackson: University Press of Mississippi, 1996.

Williamson, James A. *The Voyages of the Cabots and the English Discovery of North America under Henry VII and Henry VIII*. London: Argonaut, 1929.

Learning from Rivers

Toward a Relational View of the Anthropocene

ALLISON NOWAK SHELTON

Jason M. Kelly, Philip Scarpino, Helen Berry, James Syvitski, and Michel Meybeck, eds. ***Rivers of the Anthropocene***. Berkeley: University of California Press, 2017.

Martin Knoll, Uwe Lübken, and Dieter Schott, eds. ***Rivers Lost, Rivers Regained: Rethinking City-River Relations***. Pittsburgh, PA: University of Pittsburgh Press, 2017.

Ellen Wohl. ***A World of Rivers: Environmental Change on Ten of the World's Great Rivers***. Chicago: University of Chicago Press, 2011.

In the humanities, critics are reconceptualizing human history and culture in light of issues around hydrology, including climate change, water scarcity, water restoration, water resource infrastructure, and water rights. The "oceanic turn," of which this issue is part, is aimed at rethinking large bodies of water like oceans that traverse national and cultural boundaries as containing vast, submerged knowledge and history. Rather than bounded physically or historically, oceans are being acknowledged as crosshatched and networked, layered with human-nonhuman, material-discursive connections that challenge our notions of time, space, culture, history, and humanity at large. But what of rivers—those deceptively linear, bounded bodies that feed into the oceans and continually foil our efforts to control them? Rivers connect glacial melt and other headwaters to the seas, carrying nutrients, sediment, and animal life, and as such they bridge deep geological time and historical surface time. This relationship suggests connections between quantitative scientific study and qualitative, affective experience of the material environment. Fittingly, the great rivers of the earth often serve mythological and narratological functions, animating cultural histories not only as life givers but also as metaphors for our own human lifelines. Currently, rivers are at the heart of overlapping physical, political, and cultural aspects of water-rights discourses, bearing the brunt of overcrowding, pollution, industry, and climate change, and are therefore at the frontlines of concerns about the Anthropocene.

The geoscientist Ellen Wohl's book *A World of Rivers* posits river degradation as a comparative field that informs and theorizes planetary anthropogenic engagement

ENGLISH LANGUAGE NOTES
57:1, April 2019 DOI 10.1215/00138282-7309755

without overlooking site-specific struggles. She does not mention the Anthropocene per se but rather considers both ecological processes and cultural conceptualizations of rivers as vital to the future of water management, as well as hydrocriticism more broadly. The edited volumes *Rivers of the Anthropocene* and *Rivers Lost, Rivers Regained* put this conversation to work by examining relations between rivers and human settlement through the lens of history and globalization, which in turn provides a nexus for hydrologic issues. All three volumes under review regard river management as historically entangled with the narratives humans have told and continue to tell about rivers, and as such they encourage a combination of intersecting global and local approaches as key to the future of river ecosystems and, by extension, the future of the planet. A call to action emerges to transform the anthropocentrism that imagines humans as the primary beneficiaries of river resources in order to tease out interconnections between river management policies, social justice, and environmental degradation. Like other aspects of the environment, cultural and regional perspectives on rivers vacillate according to specific circumstances, creating competing demands that help define a river's meaning and value. An environmental policy enactment, a family engaged in their morning washing, a ceremonial ritual: these all exist as consequences of various material and discursive histories, and all contribute to changing a river's ecosystem. These are issues of imagination and perception as much as of ecology and management, which is suggestive of how literary studies may illuminate the power and potential of narrative river imaginaries. I briefly consider Arundhati Roy's representation of the Meenachal in her 1997 novel *The God of Small Things* at the end of this review essay.

Entangled Elements

Rivers have always been a conceptual issue. The early cradles of civilization were all located along large river valleys, and as such rivers have occupied important symbolic positions from the earliest accounts of cultural and literary history. In particular, the combined importance of river resources to settled societies with the constant, imminent threat of catastrophic flooding and/or drought has facilitated powerful contradictory imagery of rivers as both origin and destruction of human life. This dichotomy is amplified today, as most of the world's largest rivers have been polluted and/or drained to the point of catastrophe. Our reliance on rivers for the survival of settled society is massive; therefore the potential impact of their degradation is equally massive. The Ganges, for example, provides water for 500 million people across eight states in India as well as for sizable populations of Bangladesh, Nepal, and China, and it is one of the most polluted and ecologically stripped rivers in the world. The story is similar for rivers across the globe: Amazon, Nile, Danube, Mississippi, Congo, and so on. Hence river management is of the highest importance for river conservation and restoration and the future of water security, public health, and a host of other issues. But there is no clear way forward for river management, partly because geologically and geographically specific issues clash with political, socioeconomic, and cultural boundaries and notions of river functionality and meaning. However, there is no question that something needs to change, especially in terms of our methods of modification that disregard the importance of dynamism for river ecosystems.

In her introduction to *A World of Rivers* Wohl highlights this clash between the dynamic nature of rivers and the human need to control them: "People too often view constantly changing rivers as inconveniences. We try to stabilize them by confining them in single straight channels. . . . This confinement diminishes complexity and diversity of habitat that nourish abundant and varied species of plants and animals" (1). Wohl's book is an accessible foray into both the ecological processes of large rivers and their history with modification by humans. She is not an environmental historian and admits that she does not delve into the social and cultural histories of the riverine areas she surveys. Instead, she focuses only on the specific human modifications of rivers and the extreme diversity and dynamism that modification impedes. But rivers are not passive victims of the Anthropocene for Wohl. Rather, they are entangled with humans, as they are with so many biological communities: "Relative to the percentage of the landscape they occupy, river corridors host a disproportionately large number of plant and animal species. Rivers almost entirely dominate the transport of sediment to the oceans . . . [and] [b]iochemical processes along rivers and adjacent wetlands govern the amount of nitrogen, carbon, and other nutrients reaching coastal areas and thus strongly influence coastal and oceanic productivity" (4). As such, Wohl demonstrates that rivers are equally important to the planet's health as to settled society.

Rather than individual units, Wohl encourages us to think of rivers as actually making up one "round river," a concept drawn from Aldo Leopold, that emphasizes the global connections between seemingly separate ecosystems and the cycling of nutrients and energy. She does take each of the ten rivers in turn, assessing three that are, so far, less affected, the Amazon, Congo, and Mackenzie; two "undergoing rapid change," the Ganges and Chang Jiang; and five that are "heavily altered," the Ob-Irtysh, Nile, Danube, Mississippi, and Murray-Darling (2). But to emphasize the global connections between these river systems, Wohl also follows a hypothetical water droplet as it travels from one river to another through the water cycle, finding its way into oceans, evaporating and entering the atmosphere, at times becoming precipitation and even runoff. This travelogue-style journey takes place in the interludes between chapters. Far from a gimmick, the ongoing narrative is highly educational, not only explaining how the rivers are connected over time and space, and how an event in one river can affect another even a continent away, but also explaining each ecosystemic possibility along the way. For example, in the first interlude Wohl's water droplet falls into the upper basin of the Amazon, then journeys for nearly four months down the river before joining the Atlantic Ocean. She first stops to explain the distinctive isotropic signature that allows scientists to track the Amazon's water movement, then briefly explores the "infinite possibilities" for the water droplet once it has moved northward, including being recycled through reef communities, ingested by a parrotfish, "expelled" and "taken up by" plankton, and so forth (39). But in the end, our hydrologic hero is evaporated and makes its way through several atmospheric convection cells before joining a polar-front jet stream above Europe. Unfortunately, while in the jet stream it picks up a variety of toxic materials released from European cities and carries them over Siberia, where they all enter the Ob-Irtysh together as snowmelt.

This imaginative rendering of the cyclical nature of rivers, as well as the other ecosystemic processes related to and determined by rivers, helps us envision a truly relational, planetary fluvial environment, one inclusive of both human and nonhuman elements and one that, although fragmented, cannot be distilled down to those fragments in isolation. The water droplet may be hypothetical, but the "round river" is not, especially when paired with Wohl's geological analyses of the internal relationality of each of the ten rivers she considers. For example, in her chapter on the often-overlooked Mackenzie, which runs through Canada's Northwest Territories, she explains, "As it flows north to the Arctic Ocean, the Mackenzie collects sediment from western headwaters in the Rockies and water from countless ponds and lakes scattered across the plains in one of the world's largest wetland complexes" (297). In every chapter Wohl highlights the vast variety of elements created, moved, and changed within the scope of any given river by telling the "story" of the river: all the elements are characters, as is the river itself. In fact, though Wohl does not focus on human stories, the book helps retheorize the storytelling nature of our understanding of rivers and our alterations of them, in the vein of Serenella Iovino and Serpil Oppermann in their edited volume *Material Ecocriticism*. It is through all the "material forms emerging in combination with forces, agencies, other matter," Iovino and Oppermann explain, "that 'the world we inhabit,' with all its stories, is 'alive.'"[1] Indeed, the network of languages, cultures, and traditions that have contributed to modifying rivers, suggests Wohl, are no more or less important than the network of tributaries, sediment, animal life, chemical processes, and movement of rivers. All must be taken into consideration when one assesses a river.

In this way, *A World of Rivers* decenters the human while still acknowledging the human capacity to induce systemic change and put these rivers at risk. Large rivers flow through and alongside so many populations; they are transregional and transcultural, providing distinct ecosystems but with blurry boundaries. In deference to this fact, Wohl is vague on assigning agency for the rivers' modifications, though she walks her readers through the effects of levees, dams, industrial pollution, and other human endeavors.[2] It is here that two recent interdisciplinary edited volumes in the social sciences nicely complement Wohl's geoscientific survey. It is, after all, important to face how policies, cultural encounters, population rise, and industrial escalations have accelerated the rise of river modification and produced a previously unparalleled scale of degradation, especially in terms of water-rights discourse. *Rivers of the Anthropocene*, edited by Jason M. Kelly et al., and *Rivers Lost, Rivers Regained*, edited by Martin Knoll, Uwe Lübken, and Dieter Schott, tackle these social entanglements more directly. *Rivers of the Anthropocene* is primarily interested in the coconstituency of human history and rivers. As Kelly explains in his introduction, humans have been modifying rivers as long as we have been relying on them. The volume takes this anthropogenic constant as its starting point, grappling with the idea that when it comes to rivers, there is no clear "return" to a preanthropogenic, "natural" condition for which it would be prudent to strive. "After more than a century of research on rivers and their physical and biotic makeup, we still lack robust baselines as to how these freshwater ecosystems function," admit Andy Large, David Gilvear, and Eleanor Starkey (24). "This paucity of reference points," they continue, "hinders widespread understanding of what ecosystem services are delivered by rivers either as natural systems with humans disturbing them or as human systems with remnants of natural aquatic ecosystems embedded in them" (24).

This cyclical entanglement means that continued river modification is inevitable, but it is difficult to know the way forward. The volume explores narratological trends that create and define hydrologic goals and ethics, as well as ideological assumptions behind various quantitative analyses and policies. This is important work, as river governance alters livelihoods and tends to divide people, sometimes literally, determining access and appropriate use of water[3] as well as the methods of river modification, which often displace and/or put at risk large populations of disenfranchised communities. A number of contributors bring up the problematic lens of the Anthropocene itself that frames the degradation and/or regeneration of waterways as a species issue, ignoring uneven and unequal contributions by and risks to various societies. Says Sina Marx:

> Current disaster research points out that marginal groups are more vulnerable to disruptions. . . . Any means taken to mitigate possible impacts of climate change and resultant extreme events have to effectively include those most vulnerable groups. Otherwise, existing inequalities within our "species" are likely to increase to the detriment of those who have contributed little to the sociogenic changes that the Anthropocene brings about. (53)

Overall, the volume maintains that determining responsibilities to water sources and water issues requires studying what Richard Scarpino calls the "historical interplay between people and rivers," sizing up both historical approaches to rivers and their unintended consequences (113).

Here we return to the problem of conceptualization; rivers must be reconceptualized to challenge current aggressive and harmful methods of modification in terms of *both* environmental degradation *and* environmental justice. *Rivers Lost, Rivers Regained* is a collection of case studies in chapter form that offer this kind of reconceptualization in action. The analyses, which span the United States, Latin America, Europe, Africa, and Asia, show that the demands of development and urbanization have forced rivers to evolve in creative ways, yet those river transformations have then renewed the cultures and identities of the cities through which they run. Leaving aside the question of the Anthropocene, the authors simply make a case for city-river relations as being integral to the question of human-river relations. *Relation* is the operative word here, since, as I have mentioned, there is no going "back" for most of these rivers, yet the way forward is beset with clashing perspectives and interests. Fittingly, the volume is deliberately disjointed; the chapters jump around in both time and space in their level of specificity, using the interlinking of culture and ecology as their only overarching theme. In the words of Knoll, Lübken, and Schott, the case studies tease out "the entanglements between the peculiar rhythm of a river (as expressed by the irregular sequence of droughts and flooding, for example), the unique spaces and places that rivers create (by morphological changes, sedimentation, and erosion), and the multitudinous human interventions into these natural dynamics" (7). Though the subject is urban development, the authors do not overprivilege cities in isolation; Knoll, Lübken, and Schott assert that rivers exude a "fundamental influence on the spatial distribution of different types of agricultural and forestal production near urban centers," and the relationship between cities and what they call "their respective hinterlands," facilitated by river modification, "is of utmost importance not just for the city but for the hinterland itself" (5).

A Rushing, Rolling River-Sense

The three volumes reviewed suggest that both interdisciplinarity and reconceptualization are necessary if we are to solve current river crises. This means not only that examination and analysis of the state of the world's rivers should cross academic boundaries so that knowledge and methods are shared and ideological assumptions are challenged, but also that various kinds of knowledge are necessary to envision the way forward. Literature offers an avenue for both broadening and deepening analysis of river imaginaries, allowing us to theorize river symbologies in light of pressing water conditions. Roy's novel *The God of Small Things* parallels a family in decline with the shrinking Meenachal and provides powerful imagery of the interconnectivity of river ecology, politics, and history. The river is depicted across the novel's multiple timelines as both context—in the sense of embodying an environmental network of natural imagery and the ways that environment changes during the life of the principal character, Rahel—and character with agency itself. Perhaps as a foreshadowing of Roy's activist work against the Sardar Sarovar Dam project on behalf of lower-caste and tribal Narmada River residents, the Meenachal, an important feature of the Ayemenem landscape and of Roy's own childhood home in Kerala, acts in the novel as a border between the upper-caste Ammu's Syrian Christian family and the untouchable Paravan Velutha Pappen's family. When Ammu and Velutha begin crossing the river to have sex, breaking the region's age-old Love Laws, the river brings together, in Divya Anand's words, "the touchable and untouchable worlds."[4] The river is thus both boundary and connector, much like a pivotal node in a network of interactions. But it is also notably polluted; when Rahel returns to Ayemenem as an adult, the river smells "of shit, and pesticides bought with World Bank loans." She observes that a barrage (dam) built downstream has "choked" the river with oversalination and weeds. "Once it had the power to evoke fear. To change lives. But now its teeth were drawn, its spirit spent."[5]

This ecological change in the river helps enact the novel's tension between stability and change, poignantly reflected in the family's Ayemenem house. In the earlier timeline, when Rahel is a child, the house is almost indistinguishable from the river landscape, while in the later timeline, even when the dwindling flow has withdrawn from view, Rahel notes that the house "still had a river-sense. A rushing, rolling, fishswimming sense."[6] In this case, the river's movement is the setting for a loss of innocence, in the sense of both social transgression in the past and familial decline in the present. Yet Velutha's brother Kuttappen also describes the river as a woman with whom people should not trifle. She pretends to be harmless, he says, like "a little old churchgoing amooma, quiet and clean. . . . Minding her own business." But in reality, he reveals, she is wild, potentially destructive. "And minds *other* people's business," adds the younger Rahel.[7] Thus the river as boundary and connector, limit and crossing, at once relates to its simultaneity as setting and character, images of spatiality and temporality, respectively.

Reorienting the Meenachal as context and character also reacknowledges Sophie Mol's death, an incident that happens almost concurrently (despite the narrative's confusing chronology) with the outing of Velutha and Ammu and that prompts Velutha's brutal arrest. No one is really to blame for the Anglo-Indian girl's death; it is described in hauntingly peaceful terms: "There was no storm-music. No whirlpool spun up from the inky depths of the Meenachal. No shark supervised the tragedy. Just a quiet handing-over ceremony. A boat spilling its cargo. A river accepting the offering. One small life. A brief sunbeam. With a silver thimble clenched for luck in its little fist."[8] The river has no real agenda in this description, yet this "offering" sets everything else in the novel in motion, especially the loss of innocence and the fatal beating of Velutha, the titular "God of Small Things." But the Meenachal as setting and character at once can thus be read as neither Big Thing nor Small Thing but rather as a conduit between them, both witness and actant in the ecological, social, and historical overlaps of the Ayemenem environment. In other words, the river can be seen to be boundary and possibility at once, a space of human-nonhuman connection that can both welcome and destroy, enabling the novel's nonlinear narrative movement.

As a methodology, rethinking rivers in terms of their relational ecology, as well as their connections to oceanic discourse and to material and cultural history, allows for new readings of literary river imagery. We can rethink Roy's representation of the Meenachal as a conduit linking nature to culture, big picture to small, yet occupying both. Every river requires specific analysis of historical, cultural, political, and ecological specificities for the steps toward its restoration to be laid bare, but there are ways to resituate river knowledge that can decentralize human values more broadly, especially in the face of large-scale, industrial transformations of environmental landscapes that overprivilege humanity to the extreme. Instead, culture and cultural history, as well as political and economic approaches to rivers, are threads of a complex environmental network that determines and is determined by river ecology. These are ethical concerns. The way we approach our environment has everything to do with how it has been culturally conceived; therefore to balance considerations of both anthropogenic and nonanthropogenic shifts that affect the flow of rivers on the earth is to acknowledge that water can challenge our definitions of the world, and even the category of the human itself. The literature assessed in this review illustrates that the way forward is a conceptual issue, and river imagery can help us acknowledge that we, like rivers, are relational and exist on both microscopic and macroscopic levels. We are always already connected with the big and small elements of the nonhuman environment.

ALLISON NOWAK SHELTON is a doctoral candidate in English at the University of Colorado, Boulder. Her dissertation examines environmental relationality as a critical concept that illuminates ethical relationships across local, national, and international scales in the context of contemporary Indian English literature.

Notes

1 Iovino and Oppermann, *Material Ecocriticism*, 1.

2 Briefly, levees built to prevent flooding raise the sedimentation as well as the depth of flow in a river by restricting the area of flow. This creates enormous pressure on the levees, increasing the need for constant maintenance. When levees inevitably fail, the subsequent flood is much higher and more forceful than without the levees (Wohl mentions this in relation to the Chang Jiang in chap. 10). Similarly, hydroelectric dams, built to generate energy, clean water storage, and downstream irrigation, have a limited shelf life and cause innumerable, often unpredictable ecological side effects, including rising salinity that ruins nearby agriculture, widespread toxic algae growth in reservoirs that kills fish and renders stored water unsafe, increased surface-area evaporation, sinking deltas that kill floodplain vegetation and increase flood risks, and dramatic changes in river chemical composition that lead to deforestation. All of these effects tend to cause disproportionate harm to disenfranchised communities. For more, see Pottinger, "Wrong Climate for Big Dams"; and Winsemius et al, "Disaster Risk, Climate Change, and Poverty."

3 There is a documented consensus in the international community affirming "the human right to water," but that right is not consistently upheld due to the ambiguity of responsibility. For more on water rights, see Gupta, Ahlers, and Ahmed, "Human Right to Water."

4 Anand, "Inhabiting the Space of Literature," 102.

5 Roy, *God of Small Things*, 13, 118, 119.

6 Roy, *God of Small Things*, 30.

7 Roy, *God of Small Things*, 201.

8 Roy, *God of Small Things*, 277.

Works Cited

Anand, Divya. "Inhabiting the Space of Literature: An Ecocritical Study of Arundhati Roy's *God of Small Things* and O. V. Vijayan's *The Legends of Khasak*." *Interdisciplinary Studies in Literature and Environment* 12, no. 2 (2005): 95–108.

Gupta, Joyeeta, Rhodante Ahlers, and Lawal Ahmed. "The Human Right to Water: Moving towards Consensus in a Fragmented World." *Review of European Community and International Environmental Law* 9, no. 3 (2010): 294–305.

Iovino, Serenella, and Serpil Oppermann. *Material Ecocriticism*. Bloomington: Indiana University Press, 2014.

Pottinger, Lori. "The Wrong Climate for Big Dams: Why Africa Should Shun Hydropower Megaprojects." *World Rivers Review* 24, no. 4 (2009): 6–7, 15.

Roy, Arundhati. *The God of Small Things*. New York: Random House, 1997.

Winsemius, Hessel C., Brenden Jongman, Ted I. E. Veldkamp, Stephane Hallegatte, Mook Bangalore, and Philip J. Ward. "Disaster Risk, Climate Change, and Poverty: Assessing the Global Exposure of Poor People to Floods and Droughts." *Environment and Development Economics* 23, no. 3 (2018): 328–48.